Part I

Introduction

1

The Use of Statistics

Throughout this text our primary task has been to strive for clarity of exposition. The tools and techniques presented in this book have demonstrated their utility in a wide range of disciplines, but we have attempted to pull together into a single volume those procedures which have the greatest application to individuals working in food science or the food industry.

STATISTICS DEFINED

Statistics as a discipline encompasses all of those techniques which utilize numerical values to describe, analyze, or interpret experiments and experiences. Tables and charts are valid, albeit simple, statistical tools. Various descriptive values (means, standard deviations, index numbers, etc.) represent somewhat more complex tools. Statistical techniques which assist decision making when facing uncertainty (regression analysis, the analysis of variance, the Chi-square test, etc.) represent the culmination of considerable mathematical theory and applied experience. It is this last body of ideas that normally comes to mind when a scientist speaks of using statistics. This book presents all three aspects of statistics. Primary emphasis is on the last because it is the most difficult to master and also because it yields the greatest payoff as a result of the acquired knowledge.

DESCRIPTIVE STATISTICS

Descriptive statistics are normally one of the first steps toward analyzing a problem or making a decision. No individual can keep track of the movement of the prices associated with the more than 1000 stocks traded on the New York Stock Exchange. The Dow Jones Index of 30 industrial stocks provides a summary number which is designed to provide *in a single number* a description of the overall behavior of the market. Similarly, if 50 trials of a specific cooking time-temperature experiment are run, it may be sufficient to report only the average tenderness (i.e., the arithmetic mean of the 50 trials). In other words *a single value* is reported that summarizes the results in a way that can be comprehended more readily than reporting all

3

50 individual results. In general, this is the nature of descriptive statistics. They provide a means of reducing a large number of complex data points into a smaller number of summary values. Admittedly, information is lost in this process but the gain in understanding is substantial. Chapters 2, 17, and 18 deal with those descriptive statistics commonly encountered in the food industry.

STATISTICAL INFERENCE

One of the more important contributions that statistics has made to our thinking is the development of statistical inference. It would not be possible to test every can of food produced by a plant to determine whether or not net weight specifications are being met since the test procedure destroys the product. Inferential statistics provide a technique for examining a sample of the total production and utilizing that information to *infer* something about the characteristics of the total population from which the sample was drawn. Virtually everyone has drawn samples of grain or other products to determine grade or other characteristics. Few people realize the limitations that exist in the use of inferential statistics. One of the goals of this book (in addition to presenting the techniques and their uses) is to help the reader develop a healthy respect for the potential abuses that exist when dealing with statistical inference. Chapter 3 provides the basic concepts associated with the probability theory that underlies all statistical inference. Chapter 4 examines the use of the normal distribution and presents a number of simple tests. Together, these two chapters provide the nucleus of statistical thought for decision making.

STATISTICS AND THE SCIENTIFIC METHOD

There are probably as many definitions of the scientific method as there are scientists. Furthermore, scientists do not have any unique claim to the scientific method since nearly any decision maker uses some variant of the scientific method when examining problems. For our purposes, the scientific method can be considered to contain the following elements:

(1) Statement of the problem
(2) Organization of ideas and theories
(3) Statement of a hypothesis
(4) Testing the stated hypothesis
(5) Confirmation or denial of the stated hypothesis leading to a decision. The decision may result in action or may lead to the refinement or restatement of initial ideas and theories and the construction of a subsequent hypothesis.

The individual's discipline or job provides the basis for steps (1), (2), and (3). Statistics plays a role in step (3) by assisting in formulating a testable hypothesis. Statistics also carries the bulk of the load in step (4). In particu-

lar, the principles of experimental design (presented in Chapter 6 and reexamined in Chapters 7, 8, 9, and 10) provide an important tool which greatly facilitates organized thinking and problem approach. The final step (confirming or denying the hypothesis) rests on the concepts of probability presented in Chapters 3 and 4 and illustrated throughout the book with a wide variety of examples.

Since the scientific method has been presented in general terms, a concrete example may assist in clarifying the basic concepts that are important.

(1) Statement of the problem:
   A load of 50,000 carrots has been delivered to the plant for· processing. The problem is to decide whether or not to accept them.

(2) Organization of ideas and theories:
   It is known that a large number of defects in the raw product result in increased handling costs and reduced profits. It is important to determine whether or not the load has an excessive number of defects where management has made a decision based on prior experience that defects of more than 1% are excessive.

(3) Statement of a hypothesis:
   The load of carrots does not contain more than 1% defectives.

(4) Testing the stated hypothesis:
   Using the material presented in this text, a sample size would be determined and a sample drawn.

(5) Confirmation or denial of the stated hypothesis leading to a decision:
   Using the material presented in this text and the specific sample drawn, either the hypothesis would be accepted and the load of carrots used or the hypothesis would be rejected and the load of carrots returned to the vendor.

In this case, the scientific method results in a single set of actions and a final decision. In other instances, the result may be further investigation and a retracing of steps.

STATISTICS AND DECISION MAKING

It is clear from the previous section that statistics do not make decisions for people. Statistics as a discipline or a body of techniques aid the individual in organizing ideas and defining problems. They go even further by yielding important information about uncertainty as an additional guide to decision making. In the final analysis, the individual is responsible for the decision.

2

Observations

Statistics by its very nature deals with numbers which report the results of various observations. We may be dealing with the results of planned experiments of a specific research effort; alternatively, we may be collecting data through mailed questionnaires. We may be recording laboratory data such as spoilage of a product at various temperatures, or our observations may be of an in-plant nature where we may be sampling to determine product net weight. On a broader focus we may be looking at company sales, percentage of market, etc. In any event, the task of statistics is to help organize these observations into a form that is comprehensible and aids in decision making.

DATA VERSUS INFORMATION

The terms data and information are closely related. However, an important distinction exists. It is easy for the individual to get lost in the data. Our goal is to transform data into information. In its simplest form, this may only require the construction of a simple bar chart that provides some notion of how the data points are dispersed. Alternatively, the transformation from data to information may be as simple as plotting the points on a simple graph of X,Y coordinates in order to develop a feel of how X and Y move relative to one another. These notions are so simple and so commonly used that the individual does not realize what a giant step this represents in moving from data to information.

However, our goal is more ambitious than just the reduction of a large number of data points to a smaller number of summary values. With the availability of computers, the individual can get lost in the analytical results almost as easily as he could get lost in the original data points 20 years ago. It seems that few tools are more dangerous in the hands of a novice than the computer. A computer, properly used, can be of considerable aid in extracting information from data. A computer that is misused merely buries the individual in a new body of data which may be more difficult to understand than the original data points.

SOURCES OF DATA

During the California Gold Rush of 1849, a common piece of advice given freely to all new arrivals to the state was: "Gold is where you find it!" In a similar manner, it can be said that: "Data are where you find them." However, there are two broad categories of data (*viz.*, primary and secondary), and the individual must look in different places depending upon the type of data that are sought.

Primary Data

Primary data are those which the individual collects himself. There is no distinction among reporting the observations of feeding trials, measuring product color in the processing plant, recording observed defects of raw material received, observing consumer behavior in a retail store, using a questionnaire to determine consumer acceptance, or reporting the results of a taste panel. In each case, the researcher is on hand when the data are generated and recorded. He has the opportunity to design the experiment in such a way that the data collected will yield the information that is desired.

Secondary Data

In contrast to primary data, secondary data are published or recorded data. In other words, the researcher has no control over the gathering and recording of the data. Secondary data are someone else's primary data. If another firm runs an experiment and shares the data generated, for them they are primary data, for you they are secondary data. Ordinarily, secondary data will be those published in professional journals or government reports. There is an immense reservoir of data available through government reports. Production and consumption of nearly every agricultural product are reported. In addition, population, income, and price statistics are available. Appendix A.1 lists a few of the many publications that exist which provide a rich store of secondary data available for analysis when dealing with specific types of problems.

VARIABLES

The term variable has become virtually all-inclusive in statistics. In a very narrow sense it refers to something being observed that exhibits variation. The variation may be in part due to errors of measure, the innate variability of the material, or planned variation that is imposed exogenously. Thus, we speak of time and temperature as variables although they are usually variables that are controlled. That is, changes in time and temperature will normally be chosen by the researcher within the framework of an experimental design and, consequently, changed in some pre-planned pattern. On the other hand, the rate of reaction and the products of reaction are also referred to as variables. However, it is clear from experi-

ence that these variables are the result or the yield of the experiment. A repeat of a specific time-temperature combination will normally result in slightly different results (e.g., rates of reaction). That is, yield variables exhibit a certain randomness in their behavior. Just being able to divide variables into those that can be controlled and act upon the system (i.e., independent variables) and those that are the result or yield (i.e., dependent variables) is an important step toward understanding the problem being analyzed and the concomitant use of statistics.

FREQUENCY DISTRIBUTIONS

In many instances, the first step in a statistical analysis is to organize the data into the form of a frequency distribution. Table 2.1 reports the results of a questionnaire concerning milk consumption for 50 families during the first full week of April. Although Table 2.1 reports all of the data collected, few individuals can simply read Table 2.1 and conclude anything about household consumption. However, dividing households into groups based on consumption and recording the number of families in each classification as reported in Table 2.2 provides information which is more readily understood by the reader. Table 2.2 provides a feel for the number of families that are light, medium, and heavy consumers of milk. Figure 2.1, which is a simple bar chart or histogram, provides an even better means of conveying the information that was presented in Table 2.1 as unorganized data.

TABLE 2.1.   CONSUMPTION OF MILK BY 50 FAMILIES DURING THE FIRST WEEK OF APRIL (QUARTS)

28	7	15	9	28
11	12	23	4	17
5	18	25	14	19
30	21	18	23	22
17	18	13	12	11
8	6	27	24	14
15	1	16	6	16
6	13	17	26	12
20	2	10	19	21
5	9	5	14	10

See Appendix for metric conversions.

TABLE 2.2.   CONSUMPTION OF MILK BY 50 FAMILIES DURING THE FIRST WEEK OF APRIL: A FREQUENCY DISTRIBUTION

Number of Quarts Consumed	Number of Families
1– 5	6
6–10	9
11–15	12
16–20	11
21–25	8
26–30	4

See Appendix for metric conversions.

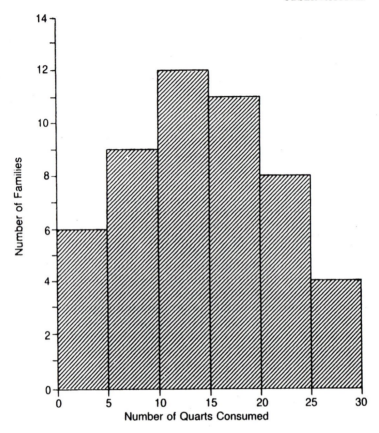

FIG. 2.1. CONSUMPTION OF MILK BY 50 FAMILIES DURING THE FIRST
WEEK OF APRIL: A HISTOGRAM

It takes time and experience to decide how many classifications to use
when organizing data into a frequency distribution. In general, fewer than
5 classifications hides more than it shows. At the other extreme, it is rare
that more than 20 classifications are needed.

In order to clarify the steps that were just taken, the following summary
is provided:

(1) The original data of Table 2.1 were considered to be incomprehensible
   in an unorganized form.

(2) It was arbitrarily decided to organize the data into 6 classes to develop
   a frequency distribution. (The choice of 6 classes was made primarily
   to facilitate the definition of a reasonable class interval since the
   choice of 10 classes would have resulted in a class interval of 3 which
   is often awkward as an interval.)

(3) Table 2.2 was constructed by counting the number (i.e., the frequency) of families that fell into each class. For Table 2.2:
For the first class:
   (a) The class interval is 5
   (b) The class limit is 1 to 5
   (c) The frequency is 6 families
   (d) The class midpoint is 3
For the second class:
   (a) The class interval is 5
   (b) The class limit is 6 to 10
   (c) The frequency is 9 families
   (d) The class midpoint is 8
Etc.

Class limits should be defined in such a way that no gaps exist. Since Table 2.1 reported only integer values, our task was simple. However, if fractional units had existed, we would have defined class limits which would have permitted the classification of the original data into mutually exclusive class intervals without intervening gaps.

A bar chart or histogram is simply a visual presentation of a frequency table. However, as shown in Fig. 2.2, the differences that may be encountered can be striking. Figure 2.2A shows the type of distribution that one ordinarily expects to find with the largest number of observations in the middle and fewer observations at the extremes of high and low consumption. Figure 2.2B shows a uniform distribution with families falling into the various classes in approximately equal numbers. Figure 2.2C shows a situation where nearly all of the families fall into a single class. Figure 2.2D shows a bimodal or cat's head distribution with large numbers at the upper and lower extremes but relatively few in the middle. Data which exhibit bimodal frequency distributions usually are caused by the pooling of information from two distinct populations, a subject which is dealt with in Chapter 4.

DESCRIPTIVE STATISTICS

Although the benefits of plotting the frequency distribution in a histogram should never be underestimated, it is only the first step toward transforming raw data into usable information. There are a number of simple descriptive statistics which can be calculated from the raw data to yield valuable information, also.

Arithmetic Mean

The arithmetic mean is commonly used in everyday life. When the ordinary individual speaks of calculating an average, he is referring to an arithmetic mean. However, statistics recognizes other types of averages which are also discussed in the chapter (*viz.*, geometric mean, median, and mode).

STATISTICAL METHODS FOR
FOOD AND AGRICULTURE

Contents

the individual to learn and develop on his own without outside assistance. The provision of exercises at the end of each chapter permits its use as a textbook as well.

We gratefully acknowledge the encouragement and assistance received from many sources: To the food and agricultural organizations who presented the problems to be solved; to our students who helped in the solution of the problems, with special thanks to Susan Epstein for working through many of the problems during the writing of this book; to the late Dr. Donald K. Tressler for the use of some material published elsewhere in AVI publications; to Dr. N.W. Desrosier and Barbara J. Flouton of the AVI Publishing Co. for their encouragement and assistance in bringing this book into being; to Drs. B.V. Lessley, V.J. Norton, R.F. Davis, and B.A. Twigg, University of Maryland, for administrative support; to Janet Mitchell, Margaret Kemph, and Brenda Derflinger for secretarial assistance and for deciphering occasionally illegible handwriting; to Barbara Ahern for the illustrations; and last, but by no means least, to Christine, Sue, and Diana.

FILMORE BENDER
LARRY DOUGLASS
AMIHUD KRAMER

September 1981

Preface

Statistics is a relatively recent development for science and business. Although mathematics has a history of several thousand years, statistics essentially traces its origin to the work of R.A. Fisher in England during the 1930s. There are earlier bits and pieces, but the onrush of statistical analysis which now surrounds us encompasses a period of less than 50 years.

Statistics is basically a tool to facilitate decision-making. Under different sets of circumstances, different perspectives are required. This book attempts to meet the two primary statistical needs faced by individuals in the food and agricultural industries: that is, statistics as an aid in research and statistics as an aid in business. The book is divided into 4 major sections.

Part I provides an introduction to the uses of statistics today, including basic concepts and definitions.

Part II examines the statistical needs of the food researcher. Here, the emphasis is on design of planned experiments, the analysis of data generated by planned experiments, and decision-making in a research environment.

Part III deals with statistical procedures that have a wide range of uses in both business and research situations. These techniques are used equally by both the researcher and the business analyst.

Part IV is concerned with those statistical methods which have primarily a business application. These procedures do not always possess the mathematical refinement of the procedures considered in Parts I, II, and III; however, they have demonstrated their usefulness to businessmen and decision-makers in thousands of applications. Their inclusion in this book is due to their widespread use and usefulness.

This book is aimed at the needs of the food and agricultural industries in both their research and business needs. It is sufficiently detailed to enable

Food Products Press, Inc., is a subsidiary of The Haworth Press, Inc., 10 Alice Street, Binghamton, New York 13904-1580
EUROSPAN/Food Products Press, 3 Henrietta Street, London WC2E 8LU England

Library of Congress Cataloging-in-Publication Data

Bender, Filmore Edmund, 1940-
  Statistical methods for food and agriculture.

  Includes bibliographies and index.
  1. Food industry and trade — Statistical methods. 2. Agriculture — Economic
aspects — Statistical methods. I. Douglass, Larry W. II. Kramer, Amihud, 1913- . III. Title.
HD9000.4.B46   1989   630'.72   89-11633
ISBN 1-56022-000-7

STATISTICAL METHODS
FOR FOOD AND
AGRICULTURE

Filmore E. Bender, Ph.D.
Professor of Agricultural Economics

Larry W. Douglass, Ph.D.
Associate Professor of Dairy Science

Amihud Kramer, Ph.D.
Professor Emeritus, Food Science

University of Maryland

Food Products Press
New York • London

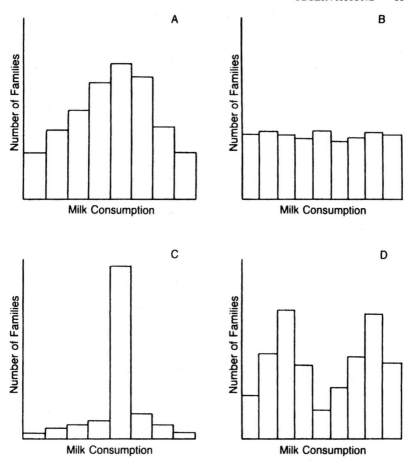

FIG. 2.2. FOUR HISTOGRAMS SHOWING POSSIBLE FREQUENCY DISTRIBUTIONS
THAT MIGHT BE ENCOUNTERED

The arithmetic mean is calculated by adding up all of the observations
and dividing by the number of observations. For the 50 observations in
Table 2.1, the arithmetic mean is:[1]

$$\frac{742}{50} = 14.84$$

To simplify instructions for statistical calculations, statisticians have
developed a variety of shorthand notations. Although considerable effort
must be expended in order to learn this notation initially, it greatly simpli-

fies communication and becomes a useful tool as an aid to analytical thinking. As an illustration, the formula for the arithmetic mean would be written as:[2]

$$\overline{Y} = \frac{\sum\limits_{i=1}^{n} Y_i}{n} \qquad (2.1)$$

Equation 2.1 would be read as Y-bar (i.e., the mean) is equal to the sum signified by the Greek letter sigma) of Y sub i, where i goes from 1 to n, divided by n (i.e., the number of observations).

The use of subscripts and summation signs results in a very compact and powerful language. For example,

$$\sum\limits_{i=1}^{2} Y_i = Y_1 + Y_2$$

which means the same thing on both sides of the equality, does not represent any saving of space. But, the expression

$$\sum\limits_{i=1}^{6} Y_i = Y_1 + Y_2 + Y_3 + Y_4 + Y_5 + Y_6$$

shows that the left hand side of the equation is still very compact but the right hand side which means exactly the same thing takes a great amount of space.

---

[1] Two questions that often cause concern in statistics are the questions of significant digits and rounding. In general, it is not possible to derive a number with more significant digits than the original data possessed. Since our original data contained only integers, it would have been perfectly correct to report the mean as 15 (i.e., 14.84 rounded to the nearest integer). Since it is common to report averages with more digits than the original data possess, we followed conventional practice and reported a mean of 14.84 although it really offers no more information than stating that the mean is 14.8 or even 15.

In statistics, the standard practice is to round to even. That is, if the digit dropped due to rounding is greater than 5, we round up, if the digit is less than 5, we round down. If the digit is exactly 5, we round up or down such that the rounded number is even, and we must look at succeeding digits, if any. For example, 4.51 rounds up to 5; 4.49 rounds down to 4; 4.50 rounds down to 4; but 5.50 rounds up to 6. This avoids any systematic bias in the rounded values.

[2] Until recently most texts in statistics used X to designate the variable under consideration. However, the use of X as the yield variable in the analysis of variance coupled with the use of Y as the yield variable and X as the independent or causal variable in regression analysis often causes considerable confusion. Many authors are beginning to use Y to designate a yield variable consistently throughout a text. We have adopted this practice in the hope that it will facilitate the learning and using of statistics.

If the saving of space on the page represented the only gain from using subscripts and summation signs, undoubtedly statisticians would have used them for that reason alone. However, the gain is greater than that. As will become apparent in subsequent chapters, the use of subscripts becomes an important tool for keeping variables, observations, and even concepts straight as problems become more complicated.

If the 50 observations of Table 2.1 are viewed as a single array of numbers, then we can say that the first column reports observations 1 through 10; the second column reports observations 11 through 20; the third column reports observations 21 through 30; etc. We can use subscript notation to define each column total.

The total for the first column $= \sum_{i=1}^{10} Y_i$

which is read as the sum of Y sub i where i goes from 1 to 10.

The total for the second column $= \sum_{i=11}^{20} Y_i$

which is read as the sum of Y sub i where i goes from 11 to 20.

The total for the fifth column would be written as

$$\text{Total}_{\text{fifth column}} = \sum_{i=41}^{50} Y_i$$

The total for all of the 50 observations would be written as:

$$\text{Total} = \sum_{i=1}^{50} Y_i$$

Thus the general formula for the arithmetic mean (i.e., Equation 2.1) would be rewritten for the data of Table 2.1 as

$$\bar{Y} = \frac{\sum_{i=1}^{50} Y_i}{50} = \frac{(Y_1 + Y_2 + \cdots + Y_{50})}{50} = \frac{(28 + 11 + \cdots + 10)}{50} = \frac{742}{50} = 14.84$$

The arithmetic mean is deceptively simple. It is easy to calculate and widely used. By and large, because of its widespread use, it is relatively well understood for most uses. However, the arithmetic mean has a number of important statistical properties that are discussed at the end of Chapter 3.

Geometric Mean

The geometric mean is another type of a statistical average. It is difficult and time consuming to calculate. For most purposes, it offers no special advantage over other types of averages. Consequently, it is rarely used. The geometric mean is defined by the following formula:

$$\text{Geometric Mean} = \sqrt[n]{Y_1 \cdot Y_2 \cdot Y_3 \cdots Y_n} \qquad (2.2)$$

That is, the geometric mean is equal to the nth root of the product of all of the observations. The usual procedure for calculating the geometric mean is to transform the data to logarithms and thereby simplify the required calculations.

Using natural logs:

$$\ln(\text{Geometric Mean}) = \frac{\ln Y_1 + \ln Y_2 + \cdots \ln Y_n}{n}$$

For the data from Table 2.1, the formula would be

$$\ln(\text{Geometric Mean}) = \frac{\ln 28 + \ln 11 + \cdots + \ln 10}{50}$$

$$= \frac{3.33220 + 2.39790 + \cdots + 2.30259}{}$$

$$= \frac{126.08978}{50}$$

$$= 2.52180$$

The geometric mean is determined by taking the antilog.

$$\text{Geometric Mean} = e^{2.52180}$$
$$= 12.45$$

Median

In contrast to the geometric mean, the median is a type of statistical average that is relatively easy to determine and offers some distinct advantages over the arithmetic mean. The median value is that value for which 50% of the observed values are larger and 50% are smaller. To illustrate with a simple example, consider the array:

$$10 \quad 17 \quad 6 \quad 15 \quad 3 \quad 9$$

We determine the median by first ordering the array in ascending values and then finding the midpoint. For our array:

$$3 \quad 6 \quad 9 \quad 10 \quad 15 \quad 17$$

If the array contains an odd number of observations, the median is the midpoint observation. Since our array has an even number of observations, the median is the arithmetic mean of the 2 middle values. In this case:

$$\text{Median} = \frac{9 + 10}{2}$$
$$= 9.5$$

In other words, for our array of 6 observations, 3 lie below the value 9.5 and 3 lie above the value 9.5.

In order to demonstrate the importance of the median, we first calculate the arithmetic mean for the same 6 observations.

$$\text{Mean} = \frac{3 + 6 + 9 + 10 + 15 + 17}{6} = \frac{60}{6}$$
$$= 10$$

In this case the mean and median are nearly equal.

Suppose that we acquire a seventh observation with a value of 290. Now our array would be:

10   17   6   15   3   9   290

The arithmetic mean of these 7 observations would be 50!—a dramatic change. In contrast, the median of the new array is 10. As a result, the median in this case would more nearly approach our general notion of "average" than would the mean. The mean is nearly 5 times as large as 6 of our 7 observations. The median provides a better insight into our data. That is, it conveys more meaningful information. Consequently, there are times when it is better to report the median than the mean. This usually occurs with a highly skewed distribution.

Skewness is a measure of the lack of symmetry in a frequency distribution. Figure 2.3 shows the relationship between the mean and median in a symmetrical and in a skewed distribution. Since income data in a society are often highly skewed, national statistics usually report median family income rather than mean family income.

For the data in Table 2.1, there are an even number of observations. When these numbers are arrayed in ascending order, the twenty-fifth value equals 14 and the twenty-sixth value equals 15. For the data in Table 2.1 the median is 14.5. That is, there are 25 observations smaller than 14.5 and 25 observations greater than 14.5.

Mode

The mode is an average in the sense that it is the value that occurs with greatest frequency. This is most easily seen in Fig. 2.3. With a small number of observations, it may be that no 2 are equal. Alternatively, there may be more than 1 mode as in the case of the data in Table 2.1 where several values occur with a frequency of 3 (viz., 12, 14, 17, and 18). However, with grouped data such as those shown in Table 2.2, we speak of the modal class rather than the mode. For Table 2.2, the modal class is the class 11–15 with 12 observations. Both the arithmetic mean (14.84) and the median (14.5) lie within the modal class.

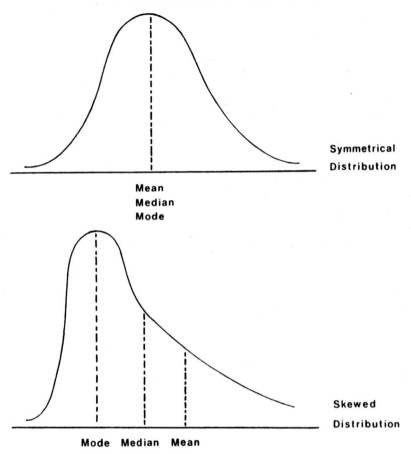

FIG. 2.3.   FREQUENCY DISTRIBUTIONS SHOWING THE RELATIONSHIP AMONG THE
MEAN, THE MEDIAN, AND THE MODE IN SYMMETRICAL AND SKEWED DISTRIBUTIONS

Percentiles

Percentiles divide an array into classifications where each group is 1% of
the total population. In other words, the first percentile is a value such that
1% of all of the observations lie below that value. In a similar manner, the
twentieth percentile is a value such that 20% of the observations lie below
it. The fiftieth percentile is the median (i.e., the value such that 50% of the
observations lie below it).

Quartiles

Quartiles are similar to percentiles except they divide the population into 4 groups. The second quartile is equal to the median. Quartiles are rarely used but surface in some publications from time to time.

Deciles

Deciles divide the population into 10 groups. The fifth decile is equal to the median. Deciles and percentiles are commonly used to report educational data and the results of standardized tests in schools.

Range

The range is the largest number minus the smallest number.

$$\text{Range} = Y_{largest} - Y_{smallest}$$

For the 50 observations in Table 2.1 the range equals 29 (i.e., $30 - 1$). The range gives a measure of the dispersion of the data.

Summary

This section has provided a brief introduction to the more common descriptive statistics. For the most part, only the mean, the median, and the range are used.

POPULATIONS AND SAMPLES

In statistics it is important to distinguish between populations and samples. The population (often called the universe) is the collection of all possible observations that might or do exist. A sample is the collection of observations actually taken. In a complete enumeration, the sample and the population are synonymous. In food science, it is rare to have a complete enumeration. Normally, we are dealing with a sample which is a subset of the total population.

The descriptive techniques discussed in this chapter apply to both samples and population. However, if we are dealing with a sample, our descriptive statistics only permit us to infer certain things about the population. Without a complete enumeration, we cannot know with certainty the descriptive statistics which apply to the total population. Chapters 3 and 4 examine this problem in more detail because this concept is the very crux of inferential statistics.

PROBLEMS

2.1   For the following data:

7	2	10	11
1	6	13	4
9	7	2	7
12	14	7	10
3	11	4	9
8	5	12	5
10	9	3	8
4	6	8	6

(a) Determine:
   (1) The arithmetic mean
   (2) The mode
   (3) The median
(b) Construct a frequency table and a histogram:
   (1) Using 5 classes with a class interval of 3
   (2) Using 7 classes with a class interval of 2

REFERENCES

FREUND, J.E. and WILLIAMS, F.J. 1958. Modern Business Statistics. Prentice-Hall, Englewood Cliffs, N.J.

KIRKBRIDE, J.W. 1975. Scope and Methods of the Statistical Reporting Service. U.S. Dep. Agric. Misc. Publ. *1308*.

SPURR, W.A. and BONINI, C.P. 1967. Statistical Analysis for Business Decisions. Richard D. Irwin, Homewood, Ill.

Suggested Further Readings

For an amusing presentation on the use and misuse of descriptive statistics see:

HUFF, D. 1954. How to Lie with Statistics. W.W. Norton, New York.

For a detailed discussion on errors in secondary data see:

MORGENSTERN, O. 1963. On the Accuracy of Economics Observations, 2nd Edition. Princeton Univ. Press, Princeton, N.J.

3

Probability

Probability theory had its beginnings with the examination of games of chance in France in the 17th century. At that time, payoffs in dice were based on experience and, as a result, some of the prevailing odds favored the gambler rather than the house. Needless to say, through the use of mathematics and probability theory this oversight has been rectified. The odds offered at modern gambling establishments are soundly based on probability theory and ensure the house percentage in the long run.

ELEMENTARY PROBABILITY

Twenty years ago it was common to hear weather reports with statements such as, ". . . a slight chance of widely scattered showers." Today, weather reports use probability statements such as, "There is an 80% chance of rain this evening, decreasing to 20% by morning." Such probability statements represent only the tip of the iceberg in that an immense amount of historical data and statistical analysis of weather patterns, barometric pressure, cloud cover, etc., must be undertaken to develop the necessary backlog of experience to be able to generate such a probability statement. This chapter begins with some simple probability distributions that are readily understood in order to develop an intuitive feel for the concept of probability. The chapter continues with some of the basic rules and definitions of probability theory. Next, the foundation of most of modern statistics (i.e., the normal distribution) is introduced and examined. Finally, mathematical expectations are presented.

COMMON PROBABILITY DISTRIBUTIONS

Probability statements are made in terms of values that range from 0 to 1. A value of 1 indicates that the outcome is certain to occur. The value of 0 indicates that the outcome is certain not to occur. The sum of the probabilities of all possible events is equal to 1.0 by definition. These statements are true whether we are dealing with continuous or discrete distributions.

Coins

A simple discrete probability event is the result of tossing a coin. There are 2 possible outcomes: heads and tails.

The probability of a head is 1/2. Using standard notation this is written as:

$$P(H) = 1/2$$

The probability of a tail is 1/2. Using standard notation:

$$P(T) = 1/2$$

Since the only 2 possible outcomes from tossing a coin are H and T, the sum of the probabilities of these events must equal 1.0.

$$
\begin{aligned}
P(H) &= 0.5 \\
+ \quad P(T) &= 0.5 \\
\hline
P \text{ (All possible outcomes)} &= 1.0
\end{aligned}
$$

This is so elementary and commonplace that it appears trivial. But we have already achieved an important grasp of probability. A few simple rules will permit us to expand this simple base into a complex analytical structure.

Theorem I The probability that 2 independent events will occur together is given by the products of their probabilities.
$P(A \text{ and } B) = P(A) \cdot P(B)$

The tossing of a coin is an experiment. The outcome of observing a head (H) is an event. Tossing the coin twice is 2 independent experiments. Observing 2 heads in a row would correspond to observing that 2 independent events occur together. The probability of this combined event is equal to the product of the probabilities of the 2 independent events. Using standard notation, we would write:

$$
\begin{aligned}
P(H,H) &= P(H) \, P(H) \\
P(H,H) &= (0.5) \, (0.5) \\
P(H,H) &= 0.25
\end{aligned}
$$

We can define every possible outcome from 2 tosses of a coin and use Theorem I to calculate the probabilities.

$$
\begin{aligned}
P(H,H) &= P(H) \, P(H) = 0.25 \\
P(H,T) &= P(H) \, P(T) = 0.25 \\
P(T,H) &= P(T) \, P(H) = 0.25 \\
P(T,T) &= P(T) \, P(T) = 0.25
\end{aligned}
$$

Since we have defined every possible outcome, the sum of the probabilities must equal 1.0.

$$P(H,H) + P(H,T) + P(T,H) + P(T,T) = 1.0$$

If we are not interested in the order of events, we can consider the possibility that our 2 tosses result in a head and a tail.

Theorem II The probability that 1 of 2 events will occur equals the sum of their probabilities minus the probability that both will occur.
$$P(A \text{ or } B) = P(A) + P(B) - P(A \text{ and } B)$$

If the events are mutually exclusive, then the probability that both will occur is 0.

Using Theorem II, the probability of 2 tosses of a coin yielding a head and a tail (regardless of order) equals:

$$P(1 \text{ Head and 1 Tail}) = P(H,T) + P(T,H) - P(H,T \text{ and } T,H)$$

However, since these are mutually exclusive events, the probability of both orders occurring is 0. As a result, our probability is:

$$P(1 \text{ Head and 1 Tail}) = 0.25 + 0.25 - 0 = 0.50$$

In a similar manner, we could calculate the probability associated with achieving at least 1 head.

$$P(\text{At least 1 Head}) = P(1 \text{ Head and 1 Tail}) + P(H,H) - P(1 \text{ Head and 1 Tail and } H,H)$$
$$P(\text{At least 1 Head}) = 0.50 + 0.25 - 0 = 0.75$$

However, for probabilities that deal with a situation where we are interested in "at least 1 event" it may be simpler to use Theorem III.

Theorem III The probability of the complement of A (i.e., all things that are not A) is equal to 1 minus the probability of A.
$$P(A') = 1 - P(A)$$

In the case of our problem of determining the probability of achieving at least 1 head, the complement would be achieving no heads (or stated another way, achieving only tails).

As a result we can write:

$$
\begin{aligned}
P(\text{At least 1 Head}) &= 1.0 - P(T,T) \\
&= 1.0 - 0.25 \\
&= 0.75
\end{aligned}
$$

These situations are so commonplace and easily comprehended that the answers are intuitively obvious. Our purpose in presenting them is to show illustrations of the basic theorems of probability in a setting where the answers are readily apparent and we can work backward to determine whether or not the theorems appear to be reasonable.

Dice

A second discrete probability distribution which is quite familiar is that which is associated with tossing a die. An ordinary die with 6 equal surfaces has the following probability distribution:

$$P(1) = 1/6$$
$$P(2) = 1/6$$
$$P(3) = 1/6$$
$$P(4) = 1/6$$
$$P(5) = 1/6$$
$$P(6) = 1/6$$

In an unbiased die, each surface has an equal chance of showing. As with all probability density functions (p.d.f.), the sum of the probabilities equals 1.0.

The tossing of a pair of dice represents 2 independent experiments, since the outcome of 1 die has no impact on the value shown on the second die. Using Theorem I, we can calculate the probability of throwing "snake eyes" (i.e., a 1 on each die).

$$P(1,1) = P(1) \, P(1)$$
$$P(1,1) = 1/6 \cdot 1/6$$
$$P(1,1) = 1/36$$

In a similar manner, the probability of throwing "boxcars" (i.e., a 6 on each die) is:

$$P(6,6) = P(6) \, P(6)$$
$$P(6,6) = 1/6 \cdot 1/6$$
$$P(6,6) = 1/36$$

To calculate the probability of these extreme events is simple because there is only a single way of achieving each event. In order to calculate the probability of throwing a "7," the use of both Theorem I and Theorem II is required.

The first step in determining the probability of throwing a "7" is to define the ways in which "7" can occur and the probability associated with each occurrence.

$$P(1,6) = P(1)\,P(6) = 1/36$$
$$P(2,5) = P(2)\,P(5) = 1/36$$
$$P(3,4) = P(3)\,P(4) = 1/36$$
$$P(4,3) = P(4)\,P(3) = 1/36$$
$$P(5,2) = P(5)\,P(2) = 1/36$$
$$P(6,1) = P(6)\,P(1) = 1/36$$

The second step is to add the probabilities of these mutually exclusive events:

$$P(1,6 \text{ or } 2,5 \text{ or } 3,4 \text{ or } 4,3 \text{ or } 5,2 \text{ or } 6,1) = 6/36 = 1/6$$

The probability of throwing an "11" can be determined in the same way. Using Theorem I:

$$P(5,6) = P(5)\,P(6) = 1/36$$
$$P(6,5) = P(6)\,P(5) = 1/36$$

Using Theorem II:

$$P(5,6 \text{ or } 6,5) = 2/36 = 1/18$$

With these probabilities and Theorem II, the probability of a "7" or "11" is:

$$P(7 \text{ or } 11) = 1/6 + 1/18 = 2/9$$

Cards

A standard deck of 52 playing cards provides another commonplace experience to explore the basic rules of probability and to add the concept of conditional probability.

A standard deck has 2 black suits (spades and clubs) and 2 red suits (diamonds and hearts). Within each suit, there exists an Ace, 2, 3, 4, 5, 6, 7, 8, 9, and 10 in addition to the 3 face cards of Jack, Queen, and King.

Since each suit possesses 13 cards, the probability of drawing a spade is:

$$P(S) = 13/52 = 1/4$$

If we replace the card and reshuffle the deck, the probability remains constant. However, if we draw cards without replacement, the probability changes depending upon the outcome of preceding draws. For example, if we draw with replacement, the probability of drawing a spade twice is:

$$P(S,S) = P(S)\,P(S)$$
$$P(S,S) = 1/4 \cdot 1/4$$
$$P(S,S) = 1/16$$

On the other hand, if we draw without replacement, the probability is:

$$P(S,S) = 13/52 \cdot 12/51$$
$$= 156/2652 = 3/51$$

The probability of drawing 2 spades in succession with replacement (1/16) is slightly more favorable than drawing 2 spades in succession without replacement (3/51). Throughout the remainder of this chapter we will deal only with problems of drawing cards from a standard deck with replacement in order to simplify our calculations.

The probability of drawing a King is:

$$P(K) = 4/52 = 1/13$$

The probability of drawing a red card is:

$$P(R) = 26/52 = 1/2$$

Using Theorem I, the probability of drawing a red King is:

$$P(\text{Red King}) = P(\text{Red and King}) = P(R)\,P(K)$$
$$P(\text{Red King}) = 1/2 \cdot 1/13 = 1/26$$

Using Theorem II, the probability of drawing either a red card or a king is:

$$P(\text{Red or King}) = P(R) + P(K) - P(\text{Red King})$$
$$= 1/2 + 1/13 - 1/26$$
$$= 14/26 = 7/13$$

Since drawing a red card is not mutually exclusive with the event of drawing a king, the probability of drawing a Red King must be subtracted out in order to avoid double counting.

The last important probability concept that we must consider is that of conditional probability.

Theorem IV: The probability that A will occur given that B does occur is equal to the probability of A and B divided by the probability of B.

$$P(A|B) = \frac{P(A \text{ and } B)}{P(B)}$$

The probability of drawing a Jack is:

$$P(J) = 4/52 = 1/13$$

The probability of drawing a face card is:

$$P(\text{Face}) = 12/52 = 3/13$$

The question that we want to address is: Given that a card is a face card, what is the probability that it is also a Jack?

$$P(\text{Jack}|\text{Face}) = \frac{P(\text{Jack and Face})}{P(\text{Face})}$$

The vertical bar is read as "given that." The equation would thus be read: The probability of a Jack given that a Face card is drawn is equal to the probability of a Jack and Face occurring divided by the probability of a Face card occurring.

Since the events Jack and Face are not independent (all Jacks are Face cards), Theorem I does not apply. The probability of a "Jack and Face" is simply the probability of a Jack or 1/13. As a result,

$$P(\text{Jack}|\text{Face}) = \frac{1/13}{3/13} = 1/3$$

Common sense agrees with this answer since one-third of all face cards are Jacks. The underlying logic of conditional probability is that the population is being redefined. In this case, instead of dealing with the entire deck of 52 cards, the conditional probability focuses attention on the 12 face cards as a newly defined population.

Everyday Events

Coins, dice, and cards have been used to illustrate the basic theorems of probability theory because the answers are readily apparent. However, the theorems presented can be applied to any situation where probability statements can be made.

If Tonja and Lisa are learning to bake cakes, the probability of a success may be less than 1.0.

Assume that for Tonja the probability of a success is:

$$P(T_s) = 3/4$$

Assume that for Lisa the probability of a success is:

$$P(L_s) = 1/3$$

The probability that both will bake successful cakes is:

$$\begin{aligned} P(2 \text{ good cakes}) &= P(T_s)\,P(L_s) \\ &= 3/4 \cdot 1/3 \\ &= 3/12 = 1/4 \end{aligned}$$

Since the probability of a failure is equal to 1.0 minus the probability of a success, then:

$$P(T_F) = 1 - 3/4$$
$$= 1/4$$
$$P(L_F) = 1 - 1/3$$
$$= 2/3$$

The probability that both cakes are failures is:

$$P(2 \text{ ruined cakes}) = P(T_F) \, P(L_F)$$
$$= 1/4 \cdot 2/3$$
$$= 2/12 = 1/6$$

The probability of at least 1 good cake is the sum of the probabilities of all of the ways of getting at least 1 good cake:

$$P(T_s, L_s) = P(T_s) \, P(L_s) = 3/4 \cdot 1/3 = 3/12$$
$$P(T_s, L_F) = P(T_s) \, P(L_F) = 3/4 \cdot 2/3 = 6/12$$
$$P(T_F, L_s) = P(T_F) \, P(L_s) = 1/4 \cdot 1/3 = 1/12$$
$$P(\text{At least 1 good cake}) = 3/12 + 6/12 + 1/12 = 5/6$$

Alternatively, we could state that:

$$P(\text{At least 1 good cake}) = 1.0 - P(2 \text{ ruined cakes})$$
$$P(\text{At least 1 good cake}) = 1.0 - 1/6 = 5/6$$

All of the probability distributions discussed have been discrete. We have done this because they are easy to manipulate. The distributions used in inferential statistics are continuous. The underlying logic is unchanged but the mathematics become considerably more complex. Fortunately, the use of tables of probabilities obviates the need for advanced mathematics. The continuous distribution basic to statistics is the normal distribution.

THE NORMAL DISTRIBUTION

The formula for the normal curve is:

$$f(Y) = \frac{1}{(2\pi\sigma^2)^{1/2}} e^{-\frac{(Y-\mu)^2}{2\sigma^2}} \qquad -\infty \leqslant Y \leqslant \infty$$

Figure 3.1 shows the general appearance of the normal curve. Fortunately, it is not necessary to work directly with the formula for the normal curve. However, a brief discussion of the equation aids in understanding this basic function.

First it is important to note that the range on Y (i.e., the random variable) is from $-\infty$ (minus infinity) to $+\infty$ (plus infinity). Although these extreme

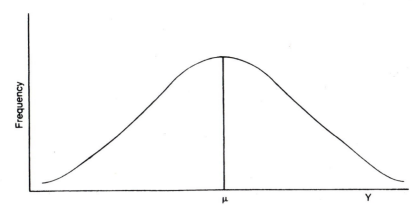

FIG. 3.1.   THE NORMAL CURVE

points are possible, Fig. 3.1 shows that such occurrences are rare. That is, observations tend to cluster around the mean (μ).

The working portion of the formula is the exponent to e, that is,

$$- \frac{(Y-\mu)^2}{2\sigma^2}$$

where μ is the population mean and σ^2 is the population variance. For any specific population, these parameters will be constants. The formula yields the relative frequency [i.e., f(Y)] for any value of Y. As Y deviates from μ, this relative frequency declines.

The constant term:

$$\frac{1}{(2\pi\sigma^2)^{1/2}}$$

simply ensures that the sum of all probabilities (i.e., the area under the curve) is 1.0.

Mean and Variance

A normal distribution is completely described by the 2 population parameters, μ (the mean) and σ^2 (the variance). The mean of a population is the center of gravity of a population. If we could completely enumerate a population, μ would be equal to the arithmetic mean. Since we normally have only a sample from a population, we calculate the statistic $\overline{Y}$ (the arithmetic mean of the sample) as an estimate of the true population mean μ.

The variance is a measure of the dispersion of the population. Figure 3.2 shows 3 normal distributions with a common mean. Figure 3.2A shows a population with a small variance; Fig. 3.2B shows a population with a somewhat larger variance; and Fig. 3.2C shows a population with a large variance.

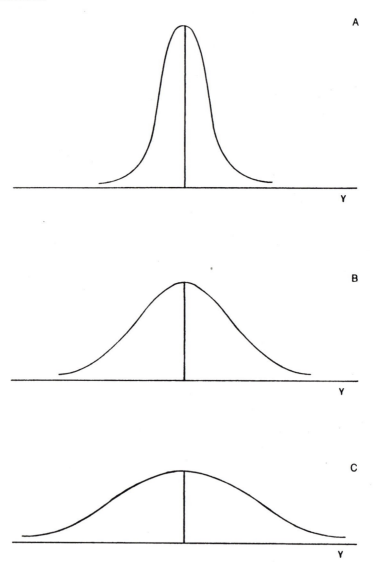

FIG. 3.2.   THREE NORMAL DISTRIBUTIONS WITH A COMMON MEAN BUT DIFFERENT VARIANCES

In spite of the fact that different populations possess different variances, the relationship between the variance and the probability of an event is constant. In other words, for any normal population, approximately 68% of the observations will lie within 1 standard deviation of the mean. (The standard deviation is simply the square root of the variance.) In addition, a little more than 95% of all observations will lie within 2 standard deviations of the mean (Fig. 3.3). As a result, if we know the population mean (μ) and variance (σ^2), we can determine whether or not a specific observation (Y_i) is a rare or common event depending on how far it is from the mean in terms of multiples of the standard deviation (the square root of the variance). In order to simplify this process, probability tables have been constructed for the normal distribution.

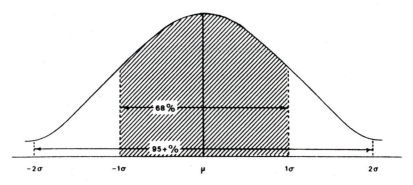

FIG. 3.3.   RELATIONSHIP BETWEEN σ AND THE PROBABILITY OF VALUES OF Y

The Use of Probability Tables

Tables have been constructed and are widely available for a standard normal deviate, that is, a variable that is normally distributed with mean zero ($\mu = 0$) and variance one ($\sigma^2 = 1$). Any normal variate with known mean and variance can be used to define a standard normal deviate by the following formula:

$$Z = \frac{Y - \mu}{\sigma}$$

If Y is normally distributed with mean μ and variance σ^2, then Z will be normally distributed with mean 0 and variance 1.0.

Suppose we are dealing with the following population:

$$\mu = 100$$
$$\sigma^2 = 25$$

and we observe the value Y $= 110$.

The question naturally arises as to whether this is a rare or a commonplace event. Our first step is to construct the appropriate standard normal deviate:

$$Z = \frac{Y-\mu}{\sigma} = \frac{110-100}{5} = 2.0$$

Using Table B.1 (in the appendix), we find the value 0.4772 associated with a Z of 2.0. The value 0.4772 is the probability that Z will lie between the mean and the value 2.0. Since one-half of the probability is associated with values below the mean, we must add 0.5000 to 0.4772 to determine the probability associated with values less than Z = 2.0. In other words, the probability that Z equals 2.0 or less is 0.9772. Or stated another way, the probability that Z would be 2.0 or greater is 0.0228 (1.0 - 0.9772). At this point we can translate this information back to our original problem.

The probability of Y being equal to 110 or greater when drawn from a normal population with mean 100 and variance 25 is 0.0228 or approximately 2%. Whether or not this is a rare event is a matter of judgement. For most people, a 2% probability is relatively small and this would be considered an unusual occurrence. Figure 3.4 shows this same information graphically.

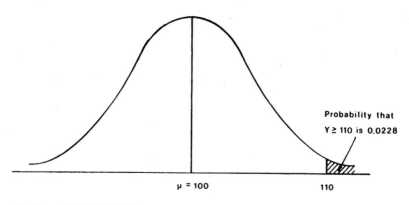

Probability that
Y ≥ 110 is 0.0228

μ = 100   110

FIG. 3.4.   PROBABILITY OF Y ≥ 110

Because the normal distribution is symmetric, usually only half of the distribution is shown in tabular form. Suppose that we are dealing with a population where

$$\mu = 1000$$
$$\sigma^2 = 100$$

and we observe Y = 993.5. Again, the question arises as to whether or not this is a rare event. Constructing the standard normal deviate:

$$Z = \frac{Y-\mu}{\sigma} = \frac{993.5-1000}{10} = \frac{-6.5}{10} = -0.65$$

Using Table B.1, row 7, column 6, the value associated with Z = +0.65 is 0.2422. In other words, the probability of Z lying between 0 and 0.65 is 0.2422. Since the distribution is symmetric, the probability that Z will lie between 0 and −0.65 is also 0.2442. The probability of a value of Z greater than −0.65 is 0.7442 (0.2442 + 0.5000). The probability of a value of Z smaller than −0.65 is 0.2558 (1.0−0.7442). At this point we can translate this information back to our original problem.

The probability of Y being 993.5 or smaller when drawn from a normal population with mean 1000 and variance 100 is 0.2558 or approximately 25%. Ordinarily, this would not be considered a rare event. Figure 3.5 shows this same information graphically.

Using the normal distribution to make decisions is discussed more fully in Chapter 4 where hypothesis testing is introduced.

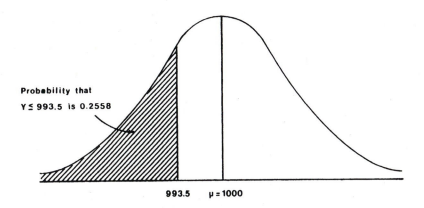

Probability that
Y ≤ 993.5 is 0.2558

993.5   μ = 1000

FIG. 3.5.   PROBABILITY OF Y ≤ 993.5

MATHEMATICAL EXPECTATIONS

A mathematical expectation is a weighted average of the values associated with all possible outcomes where the weight for each outcome is its probability of occurring. This definition is more readily understood after it has been illustrated with a few examples.

Simple Mathematical Expectations

Earlier in this chapter, the probability of achieving a head when tossing a coin was given as 1/2. If we establish a game where we are paid $1 for each head, but we must pay $1 for each tail, we can calculate the expected value of tossing a coin.

$$P(H) = 1/2$$
$$P(T) = 1/2$$
$$\text{Value of a Head} = +1$$
$$\text{Value of a Tail} = -1$$
$$\text{Expected Value of a Toss} = 1/2(1) + 1/2(-1) = 0.5 - 0.5 = 0$$

In other words, playing a game with such rules, we would expect to neither make nor lose money over a long period of play.

Of course, we can calculate the expected value associated with any other set of payoffs.

$$\text{Value of a Head} = +5$$
$$\text{Value of a Tail} = 0$$
$$\text{Expected Value of a Toss} = 1/2(5) + 1/2(0) = 2.5$$

That is, the expected value of a toss is $2.50. On any given toss, we get either $5 or nothing. Over a long period of tosses, our winnings should approximately equal $2.50 × No. of tosses. Admittedly, no one would play such a game with us unless we were willing to pay $2.50 (or more) each time we tossed the coin.

In a similar manner, we can examine the expected values associated with tossing a die. In dealing with coins, we assigned numerical values to the outcomes of a head or a tail. We did this because mathematical expectations deal with numerical values. Since the surfaces of a die possess numerical values, we can work directly with these. The expected value of tossing a die is:

$$\text{Expected Value} = 1/6(1) + 1/6(2) + 1/6(3) + 1/6(4) + 1/6(5) + 1/6(6)$$
$$= \frac{1 + 2 + 3 + 4 + 5 + 6}{6} = \frac{21}{6} = 3.5$$

The expected value (3.5) does not exist on any surface. What our mathematical expectation states is that if we toss a die repeatedly and record the outcome, over time, the average of our observations will approach 3.5.

There is no reason to restrict our values to those that exist on the surface of a die. We could assign arbitrary values.

John and Joe are tossing a die. If the result of the toss is an odd number (1, 3, or 5), Joe pays John $10. If the outcome of the toss is a 2 or 4, no money changes hands. If the result of the toss is a 6, John pays Joe $30.

The expected value of a toss for John is:

Expected Value (John) = 1/6(10) + 1/6(0) + 1/6(10) + 1/6(0) + 1/6(10) + 1/6(−30)

$$= \frac{10 + 0 + 10 + 0 + 10 - 30}{6} = 0$$

In the long run, John should expect to neither make nor lose money. For Joe the calculations are similar, but the sign of the payoffs would be reversed.

Expected Value (Joe) = 1/6(−10) + 1/6(0) + 1/6(−10) + 1/6(0) + 1/6(−10) + 1/6(30)

$$= \frac{-10 + 0 - 10 + 0 - 10 + 30}{6} = 0$$

In the long run, Joe would expect to break even in such a game.

The Use of Expectations in Statistics

In order to perform similar calculations for continuous functions, we would need to use integral calculus. However, for our purposes, we can establish a few basic rules and definitions that will enable us to handle all of the work in expectations normally required in applied statistics.

Rule 1. The expected value of a constant is the constant.
   $E(c) = c$
Rule 2. The expected value of a random variable drawn from a population is equal to the population mean.
   $E(Y) = \mu$
Rule 3. The expected value of a constant times a variable is equal to the constant times the expected value of the variable.
   $E(cY) = cE(Y)$
Rule 4. The expected value of a sum of variables is equal to the sum of the expected values.
   $E(X + Y) = E(X) + E(Y)$
Rule 5. The expected value of the product of *independent* variables is equal to the product of their expected values.
   $E(XY) = E(X) E(Y)$

In Chapter 2, we discussed the arithmetic mean and indicated that it had a number of desirable statistical properties. With the use of mathematical expectations, we can demonstrate one of those properties.

If we draw a single observation (Y) from a population, its expected value is equal to the mean (Rule 2).

   $E(Y) = \mu$

If we draw a sample of observations from a population, we calculate the sample mean $\bar{Y}$. The expected value of $\bar{Y}$ is:

$$E(\bar{Y}) = E \left(\frac{\sum\limits_{i=1}^{n} Y_i}{n} \right)$$

Using Rule 3:

$$E(\bar{Y}) = \frac{1}{n} E \left(\sum\limits_{i=1}^{n} Y_i \right)$$

Using Rule 4:

$$E(\bar{Y}) = \frac{1}{n} [E(Y_1) + E(Y_2) + \cdots + E(Y_n)]$$

Using Rule 2:

$$E(\bar{Y}) = \frac{1}{n} [\mu + \mu + \cdots + \mu]$$

Since there were n Y_i's, there are n μ's:

$$E(\bar{Y}) = \frac{1}{n} [n\mu]$$

$$E(\bar{Y}) = \mu$$

That is, the expected value of $\bar{Y}$ is the population mean. In other words, $\bar{Y}$ is an unbiased estimator of the population mean. By definition, an unbiased estimator is one whose expected value is equal to the population parameter which it estimates.

Additional properties of $\bar{Y}$ which are given without proof are:

1. $\bar{Y}$ is a least squares estimator. That is, for a given sample, the sum of squared deviations about $\bar{Y}$ is a minimum. Using standard notation:

$$\sum\limits_{i=1}^{n} (Y_i - \bar{Y})^2 = \text{a minimum}$$

In other words, if for a given sample we calculate $\bar{Y}$ and the sum of squared deviations about $\bar{Y}$, that value will be a minimum. Choosing any value other than $\bar{Y}$ will yield a larger sum of squared deviations. (This property will be understood more readily after Chapters 7–10 and 12–13 on the Analysis of Variance model and the Regression model, which are also least squares estimators, have been presented.)

2. $\bar{Y}$ is a minimum variance estimator of the population mean. That is, no other estimator of the population mean has a smaller variance (or dispersion of values).

SUMMARY

This chapter has presented the elements of probability theory and mathmatical expectations. Much more can be said on these subjects, and the interested reader may pursue these topics further through the references given at the end of this chapter. Our purpose has been to present enough material to provide a basic understanding of the concepts and an initial introduction to the notation. Our primary concern in this text is to open the doors to those statistical procedures of general use to food scientists which begin in Chapter 4.

PROBLEMS

3.1. Determine the probability of tossing a head 3 times in a row.

3.2. If a coin is tossed 4 times, what is the probability that all 4 tosses will show the same side?

3.3. In tossing a pair of dice, what is the probability of an "8"?

3.4. In tossing a pair of dice, what is the probability of throwing a 2, 3, or 12 in a single roll?

3.5. In a standard playing deck of 52 cards, what is the probability of drawing a Jack?

3.6. In a standard deck of cards, what is the probability of drawing a black Queen? What is the probability of drawing the Queen of spades?

3.7. Given that a nonface card has been drawn, what is the probability that the card is a 3? That it is a red 3? That it is the 3 of hearts?

3.8. Karl and Kurt are playing cards. If a red face card is drawn, Karl pays Kurt $26. If the Ace of spades is drawn, Kurt pays Karl $104. If any other card is drawn, no money changes hands. With each draw, the card is replaced and the deck is reshuffled. What is the expected value of the draw to Karl? Should he continue to play the game?

3.9. Show that $\Sigma(Y - \bar{Y}) = 0$.

REFERENCES

ANDERSON, R.L. and BANCROFT, T.A. 1952. Statistical Theory in Research. McGraw-Hill Book Co., New York.

CLARK, C.T. and SCHKADE, L.L. 1969. Statistical Methods for Business Decisions. South-Western Publishing Co., Cincinnati.

HARNETT, D.L. 1970. Introduction to Statistical Methods. Addison-Wesley Publishing Co., Reading, Mass.

Suggested Further Readings

For a very clear presentation of probability theory see:

LAPIN, L.L. 1978. Statistics for Modern Business Decisions, 2nd Edition. Harcourt Brace Jovanovich, New York.

For a very clear and thorough treatment of probability theory which requires some knowledge of integral calculus see:

HICKMAN, E.P. and HILTON, J.G. 1971. Probability and Statistical Analysis. Intext Educational Publishers, Scranton, Penn.

4

Sampling from a Normal Distribution

In Chapter 2, various sample statistics were introduced. We now face the task of attempting to determine how these statistics relate to the population under investigation. It is important that we understand that facts observed about a sample cannot be accepted as facts about the population from which the sample was drawn. When a particular sample is drawn, any statistic computed for that sample has only one value. However, the sample drawn represents only one of a large number of possible samples that could have been drawn, with the value of the statistic varying from sample to sample.

The problem then is to draw a sample, observe the facts about the sample, and then determine what can be said about the population. This process is known as "sample to population inference." Although population to sample inference is of great interest to the gambler and sample to sample inference may be of occasional interest to researchers, the application of statistics in food and agriculture deals largely with sample to population inference.

In the previous chapter, we introduced the subject of probability. Although several examples were given using probability in order to illustrate basic concepts, those uses have little direct application in statistics. How then is probability used in statistics? Probability is the mechanism for making statistical inference. In general, statistical inference may be classified as hypothesis testing or interval estimation. In the first case, the statistician determines the probability of observing a certain sample outcome and, based on the probability, determines if the observed sample supports or refutes a certain statement of belief. For the second case, the statistician computes lower and upper limits around a sample statistic such that the probability of its including the population parameter is known.

THE CONCEPT OF SAMPLING

Statistically, a population is the totality of individual values of a variable. The limits of the population and the variable under consideration must be clearly defined: the annual per acre (0.4 ha) yield (1 lb or 0.45 kg) of strawberries grown in Maryland, the bacterial density in the Chesapeake Bay, or the frequency of red-eyed fruit flies. If all possible values for any one of these variables are known, then the population descriptions can be made without ambiguity.

Why then do we sample? The obvious answers are time and cost. In most cases a complete population survey is too costly and the information may be of little value by the time such a survey is complete. These two considerations alone are probably sufficient to warrant the use of sampling procedures. However, the sampling process and/or the analysis may be destructive, thus dictating a sampling procedure. For example, a food processor could not permit the analysis of every package, to determine content, since he would have no product to sell. Also, do not be misled into believing that a complete population survey would necessarily give the true values. In some cases, the use of sampling may actually increase the accuracy of the resulting figures. If the analysis of a small sample permits the use of trained personnel and/or more sophisticated techniques, the sample estimates may be more accurate.

A sample is simply a part of the total population. The procedure used to choose a sample is important. Since a major purpose of sampling is to make inferences about the population, the sample must be representative of the population. The yield of strawberries in Maryland can be used to illustrate the imperative principle of representiveness. Having described the target population, we would randomly choose n (sample size will be discussed later) fields of strawberries from the whole state and measure the yield and area in each field. Choosing fields from only one county would not likely be representative of the whole state; and using fields in another state would lead to erroneous conclusions since they are outside the target population.

A *simple random sample* is defined as one in which all elements in the population have an equal and independent chance of being included in the sample. By independent, we simply mean that the selection of one element does not influence the selection of any other element in the population.

Simple random sampling is one of several sampling procedures which are generally referred to as probability sampling. The requirements of a probability sample are (1) every element of the population has a known probability of being included in the sample, (2) the sample is drawn by some method of random selection consistent with these probabilities, and (3) the probabilities of selection are taken into account in making population estimates from the sample. Samples which possess these characteristics permit the application of probability theory so that unbiased estimates of population parameters and valid estimates of sampling error can be obtained.

The probability of selection need not be equal for all elements of the population. What is necessary is that the probabilities be known. In fact, the use of sampling plans with unequal probabilities will result in increased precision for some types of populations. Three other probability sampling plans which may be of interest to some readers are stratified, cluster, and systematic random sampling.

Another consideration which should be mentioned concerns the question of sampling with or without replacement. Replacement simply refers to returning the selected item to the population after each draw, thus making it possible that an item may be included more than once in a sample. When samples are drawn from a population of infinite size, no distinction needs to be made. Also, samples drawn from populations of finite size with replacement can in effect be treated as if drawn from an infinite population. However, samples drawn without replacement from finite populations require alternative calculating procedures. The sampling procedure assumed throughout this text does not take into account this last situation; however, when sample size is less than 10% of the population size, the procedures given should still be adequate.

THE DISTRIBUTION OF SAMPLE MEANS

In this section, we present three important facts about sample means drawn at random from normally distributed populations. We are interested in the distribution of sample means because most of the techniques presented in Part II of this text (the most useful statistical techniques used in biological research) are concerned with inferences about population means. At the beginning of the next section a theorem is presented which has more general application, since it is not restricted to the normal distribution.

Let us first consider the case for the normal distribution. Given a normal distribution with mean μ, variance σ^2, and samples drawn at random, the following facts are of interest in statistics.

(1) The means of samples from a normally distributed population form a normally distributed population of means, regardless of the size of the sample.
(2) The mean of the population of means is equal to the mean of the sampled population ($\mu_{\bar{Y}} = \mu$).
(3) The variance of the population of means is 1/n of the variance of the sampled population ($\sigma_{\bar{Y}}^2 = \sigma^2/n$) and the standard deviation of the population of means is $1/n^{1/2}$ of the standard deviation of the sampled population ($\sigma_{\bar{Y}} = \sigma/n^{1/2}$).

The last quantity, the standard deviation of means, is also referred to as the standard error of a mean or more commonly, simply the *standard error*. Note that the standard error is to the population of means what the standard deviation is to the population of individual elements. Just as approximately 68% of elements lie between $\mu - \sigma$ and $\mu + \sigma$, 68% of the sample means lie between $\mu_{\bar{Y}} - \sigma_{\bar{Y}}$ and $\mu_{\bar{Y}} + \sigma_{\bar{Y}}$.

Although it is not necessary to do so in practice, it may help our understanding of the distribution of means if we visualize what would happen if we repeatedly sampled a normal population. Suppose $\mu = 20$ and $\sigma = 6$, and, further, suppose we draw all possible samples of size n = 4. Figure 4.1

illustrates this process. If we were to tabulate the sample means and form a new population of means, we would obtain a normal distribution with $\mu_Y = 20$ and $\sigma_Y = 3$. That is, drawing samples of size 4 from the population shown at the top of Fig. 4.1, and computing means of these samples would generate the population shown at the bottom of Fig. 4.1.

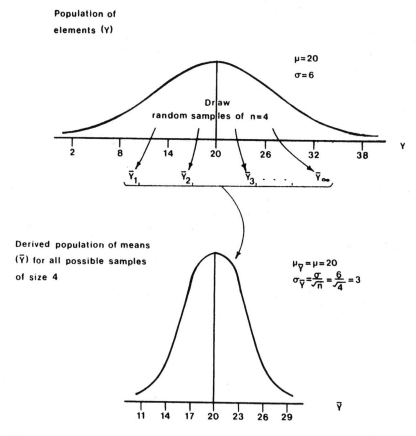

FIG. 4.1.   ILLUSTRATION OF THE SAMPLING OF THE POPULATION OF ELEMENTS (Y) TO CREATE THE POPULATION OF MEANS ($\bar{Y}$)

LARGE SAMPLE INFERENCE WITH KNOWN VARIANCE

It is easier for the student of statistics to first study the procedure based on large samples (though seldom used in practice) than it is to immediately examine the more commonly used procedures for small samples. For this reason, we will present the basic material related to hypothesis testing, assuming large samples drawn at random from a population (with known variance). This leads us to one of the more important theorems in statistics, known as the *Central Limit Theorem*, which is formally stated as follows:

Theorem—Given a population with mean μ and variance σ^2, the *distribution of sample means* drawn at random from *any population* approaches the normal distribution with *mean μ* and *variance σ^2/n* as the sample size increases.

Note that the theorem says "any population." That is, regardless of how the population is distributed, the means of samples of size n will be approximately normally distributed if n is sufficiently large.

Tests of Hypotheses

A *hypothesis* is a statement of belief, which may or may not be true about a population parameter. The *test of a hypothesis* is a comparison of the statement of belief with newly and objectively collected facts. If the newly collected facts are shown to be consistent with the stated belief, the hypothesis is accepted. If the newly collected facts do not support the statement of belief, the hypothesis is rejected.

Along with the hypothesis to be tested (null hypothesis) there will be one or two alternative hypotheses, which cannot be tested. After the null hypothesis is checked with the newly collected facts, the researcher will, if the evidence indicates he should, accept the null hypothesis. If the evidence indicates that the hypothesis should be rejected, then the researcher will adopt one of the alternative hypotheses by default.

The problem of hypothesis testing is a relatively simple problem if facts are obtained on the total population. However, as is more likely the case, the newly collected facts are based on a sample. The conclusions based on sample information are not always correct.

The hypothesis testing procedure can be thought of as consisting of three parts.

(1) Statement of the hypotheses and selection of the level of significance.
(2) Computation of the test statistic and determination of probability.
(3) Acceptance or rejection of the null hypothesis and the statement of conclusion.

Let's examine the steps and consider the case of large sample test of hypotheses regarding the population mean μ. It is known from the Central Limit Theorem that the sample mean is normally distributed with mean μ and variance σ^2/n.

Suppose that we believe that the population parameter is equal to μ_0, our hypothesized value. We may now state the null hypothesis (H_0):

H_o: $\mu = \mu_o$, that is, the true population parameter is equal to our hypothesized value.

Note that the null hypothesis is a statement of equality as is always the case in writing the null hypothesis. We believe that the population parameter is *equal* to some hypothesized value μ_o. It is the null hypothesis which is under test. The evidence to be collected will be evaluated with respect to support for or lack of support for the null hypothesis.

Two Alternative Hypotheses May Be Proposed.—

H_1: $\mu > \mu_o$, the true population value is greater than our hypothesized value *or*

H_1: $\mu < \mu_o$, the true population value is less than our hypothesized value.

If we wish to examine either one of the alternatives (but not both), then it is called a *one-tailed* test of hypotheses. When we choose to consider both of the alternative hypotheses, it is called a *two-tailed* test of hypotheses and is written as:

H_1: $\mu \neq \mu_o$, the true population value is not equal to our hypothesized value.

Next We Must Select Our Test Statistic.—The *test statistic* is a value calculated from the observed data, and is based on the null hypothesis. The sample test statistic is then evaluated to determine if the observed statistic is a rare or commonplace event given the null hypothesis. In the previous chapter we introduced the standard normal distribution. It was shown that for any observation from a normal distribution the standard normal deviate equals:

$$Z = \frac{Y - \mu}{\sigma}$$

Note that the relationship simply says that an observation's deviation from the mean $Y - \mu$, divided by the standard deviation σ, is a standard normal deviate.

The importance of the central limit theorem should now become clear to the reader. It states that sample means of size n drawn at random from any distribution are normally distributed with mean μ and variance σ^2/n. Consequently, using the procedures illustrated in Chapter 3, we can find the probability of an observed $\bar{Y}$ by computing the standard normal deviate:

$$Z = \frac{\bar{Y} - \mu}{(\sigma^2/n)^{1/2}} = \frac{\bar{Y} - \mu}{\sigma/n^{1/2}}$$

Since we wish to evaluate the sample $\bar{Y}$ relative to the null hypothesis, we compute the test statistic:

$$Z = \frac{\bar{Y} - \mu_0}{\sigma/n^{1/2}}$$

The test statistic simply tells us in standard deviation units (Z) how far the observed mean ($\bar{Y}$) is from the hypothesized mean (μ_0).

When the deviation of the observed sample mean from the hypothesized mean is small enough to be attributed to *sampling chance alone* (experimental variation), then the null hypothesis is accepted. However, if the deviation of the observed sample mean from the hypothesized mean is too great to be attributed to *sampling chance alone*, then the null hypothesis is rejected and an alternative is accepted.

To determine if the deviation of a sample mean from the hypothesized mean can be attributed to chance alone, we calculate the probability (P) of observing an outcome of the experiment as *unusual* or *more unusual* than the observed sample mean assuming the null hypothesis were true. It is the convention in hypothesis testing to use the magnitude of P as a criterion for accepting or rejecting the null hypothesis. If $P \leq \alpha$, the null hypothesis is rejected and an alternative hypothesis is accepted. If, however, $P > \alpha$, then the null hypothesis is accepted. In the hypothesis testing procedure α is referred to as the *level of significance*. The value of α is selected by the researcher in advance of testing any hypotheses. Values of 0.05 and 0.01 are commonly used although the choice is somewhat arbitrary.

Assuming that the researcher has followed the procedure described above for $\alpha = 0.05$, we are likely to see statements such as: "The null hypothesis is rejected at the 5% level," or "mean A is significantly greater than mean B at $P < 0.05$," indicating that the sample evidence was sufficient for rejection of the null hypothesis. At times authors may be very brief and simply state "the difference is significant."

Such statements are all that is necessary to complete the test of hypothesis. However, such statements of conclusion, although correct, do not provide very precise information about the test of hypothesis. Also, since the selection of the level of α is, in general, an arbitrary decision, the reader may not concur with the choice of α and is provided with too little information to permit him to reach his own conclusion. In most cases, the reader would be interested in knowing how strongly the data support or refute the hypothesis. However, it may not always be practical to report the actual probability. Instead the researcher may choose a small number of probabilities, such as 0.10, 0.05, 0.01, and 0.001, and report his conclusions using $P > 0.10, P < 0.10, P < 0.05, P < 0.01$ or $P < 0.001$. For example, the researcher might state that "mean A is significantly greater than mean B at $P < 0.001$." Such a statement will thus indicate the strength of his conclusion and provide considerably more information to the knowledgeable reader.

Let's consider some numerical examples to illustrate the hypothesis testing procedure. Suppose that a new source of a cereal grain for the manufacturing of a certain product has become available. The variance of protein content in the cereal grain is 2.25% and it is hypothesized that protein content equals 12.0%. One hundred random samples are analyzed and found to have a mean protein content of 11.8%.

Statement of Hypotheses.—

H_o: $\mu = \mu_o$
H_1: $\mu \neq \mu_o$

Since in this case $\mu_o = 12.0$

H_o: $\mu = 12.0$
H_1: $\mu \neq 12.0$

Level of Significance.—

$\alpha = 0.05$

Computation of the Test Statistic.—

$$Z = \frac{\bar{Y} - \mu_o}{(\sigma^2/n)^{1/2}} = \frac{11.8 - 12.0}{(2.25/100)^{1/2}} = -1.33$$

Determination of Probability.—

$$P = P_1(Z < -1.33) + P_2(Z > 1.33)$$

Using Appendix Table B.1 we find the probability 0.4082 in the body of the table associated with the Z value of 1.33. Since the table tabulates values between $Z = 0$ and $Z = 1.33$, the probability of a value less than 1.33 is then $(0.5000 - 0.4082) = 0.0918$. Since this is a two-tailed test and the Z distribution is symmetric, values greater than $Z = 1.33$ are also as unlikely as those for $Z < -1.33$. Consequently, the test statistic probability is $P = 0.0918 + 0.0918 = 0.1836$. This process is graphically shown in Fig. 4.2.

*Acceptance or Rejection of the Null Hypothesis.—*The sample evidence is insufficient to reject H_o since $P > \alpha$; that is, $(0.1836 > 0.05)$. The results indicate that the difference between $\bar{Y}$ and μ_o is small enough to be the result of sampling chance alone. That is, a sample deviation of 0.2 from the population mean given a standard deviation of 0.15 [from $(\sigma^2/n)^{1/2}$] would not be unexpected.

Statement of Conclusion.—"The protein content of 11.8% for the new source of cereal grain is not significantly different at the $\alpha = 0.05$ level from the hypothesized content of 12%."

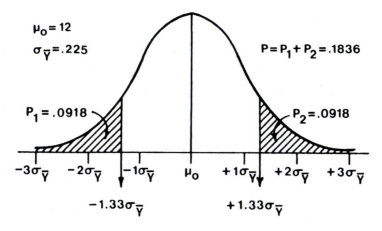

FIG. 4.2. PROBABILITIES FOR TWO-TAILED TEST OF HYPOTHESES

Note that we did not say that the protein content of the new source was 12%. Although the difference may be subtle, it is an important distinction. The hypothesis testing procedure does not prove H_o to be true, rather it fails to prove H_o false. Therefore, we accept H_o by default since sample evidence is insufficient to support rejection.

Given the results of our calculations, we could have stated our conclusions more strongly by saying that the mean was not significantly different $(P>0.10)$ or even $(P>0.18)$. However, the beginning student of statistics may feel safer simply using the common values of 0.05 and 0.01 for α. With experience the individual should develop an intuitive feel with regard to the appropriate use of alternative levels of significance.

Suppose that a certain producer is concerned with insect damage on his vegetable crop. He is told that the standard deviation in egg counts is 40 per plant and that if the number of eggs per plant exceeds 70, spraying is necessary to prevent economic losses. A random sample of 80 plants results in a mean of 79 eggs per plant.

Statement of Hypotheses.—

H_o: $\mu = 70$
H_1: $\mu > 70$

Level of Significance.—

$\alpha = 0.10$

Computation of Test Statistic.—

$$Z = \frac{79 - 70}{40/80^{1/2}} = 2.01$$

Determination of Probability.—

P = 0.0222

Since the alternative hypothesis is > 70, this is a one-tailed test and rejection of the null hypothesis occurs only if the sample evidence indicates that μ is greater than 70. Therefore, we compute the probability only for the right hand tail of the distribution (Fig. 4.3.). Again using Appendix Table B.1, we find the P (Z>2.01). In Appendix Table B.1 we find the value 0.4778 (Z = 0 to Z = 2.01) associated with a Z score of 2.01. The P (Z>2.01) then equals 0.5000 – 0.4778 = 0.0222.

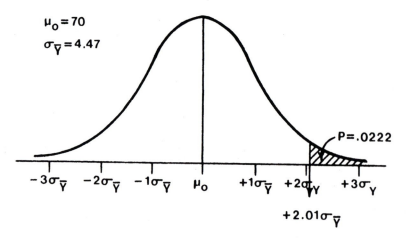

FIG. 4.3. PROBABILITY FOR ONE-TAILED TEST OF HYPOTHESES

*Statement of Conclusion.—*Reject H_o, concluding that the number of eggs per plant is greater than 70 (P < 0.10), since P < α (0.0222 < 0.10). This example differs from the previous one in several respects. It is stated *a priori* that only deviations in one direction are of interest. Therefore a one-tailed alternative hypothesis is appropriate. The only other case where a one-tail test may be conducted is when due to the nature of the response, the resulting deviation (if any) can occur in only one direction. The probability is calculated only for Z > 2.01, since this is a one-tailed test of hypothesis concerned with values >$μ_o$. Lastly, note again that we have not proven either hypothesis to be true; however, the evidence strongly refutes the null hypothesis. There is only about a 1 in 45 chance (P = 0.0222) that a value as large as the one observed (i.e., 79) could have resulted from sampling chance alone if the true mean were 70. The risk of rejecting the null hypothesis is much less than the 1 in 10 chance that the producer indicated he was willing to take by setting the level of significance at 10% ($\alpha = 0.10$).

Type I and Type II Errors

In hypothesis testing there are two types of errors which can be made.

(1) *Type I*—The rejection of a true null hypothesis.
(2) *Type II*—Acceptance of a false null hypothesis.

The probability of committing a Type I error is designated as α and the probability of committing a Type II error is designated as β.

In hypothesis testing it is desirable to keep both α and β as small as possible. Unfortunately, α and β are inversely related. If we try to minimize α, then β increases, and if we try to minimize β, then α must be greater. In general, the values of α and β should depend on the consequences of committing Type I and Type II errors, respectively. Suppose we are examining components to be used in electric heart pacers. Our null hypothesis is that the components meet some given quality standard. If the null hypothesis is false, but sample evidence leads us to accept H_0 (i.e., that they meet the standard) we have committed a Type II error. Such an error could be quite serious since the use of faulty heart pacers could result in the loss of life. Consequently β should be quite small (say 0.01 or less). On the other hand, suppose that the null hypothesis is true and we reject H_0 (conclude that the components are faulty) we have committed a Type I error. Perhaps no great harm has been done, since this action would not result in loss of life. However, such an error does result in certain costs since components which are in fact of acceptable quality will be discarded. Therefore, some protection is needed, but perhaps, considering the consequences, α can be larger than β (say 0.10 or greater). Table 4.1 summarizes the possible results of testing a hypothesis.

TABLE 4.1.   POSSIBLE OUTCOMES OF HYPOTHESIS TESTING

Null Hypothesis	Our Decision	
	Accept H_0	Reject H_0
True	Correct decision	Type I error
False	Type II error	Correct decision

Hypothesis testing is the most common application of statistics in biological research. The exact statistical test to be used varies with the data and the hypotheses to be tested. However, the general principles of hypothesis testing remain the same for all tests of hypotheses. The rest of this section will be devoted to examining in more detail how α and β are determined and what affects their values.

The probability of committing a Type I error is determined by the researcher. Before conducting a test of hypothesis, the researcher selects the level of significance (α) he wishes to use for his experiment. As mentioned earlier, many researchers simply set $\alpha = 0.05$ since it is so commonly used in biological research. If the researcher understands the consequences of both types of errors and recognizes the relationship between α and β, he may wish to select a level of α that is meaningful for his experiment.

Let's return to the hypothesis testing procedure presented in the previous section to examine these concepts. Suppose we wish to test some hypothesis about μ based on a sample of size n from a normal population with variance σ^2. Suppose that we are concerned that the mean may be greater than 50.

Given: H_0: $\mu = 50$
          H_1: $\mu > 50$
          $\alpha = 0.05$
          $\sigma^2 = 100$
          $n = 4$

This hypothesis testing procedure may be diagrammed as in Fig. 4.4. The probability of the shaded area under the curve greater than 58.2 was found using Appendix Table B.1. To determine this value we note that for a one-tailed test and $\alpha = 0.05$, the probability under the normal curve from Z to Z = (?) would equal $0.5000 - 0.0500 = 0.4500$. Searching the body of the Appendix Table B.1, we find the values 0.4495 and 0.4505. These two

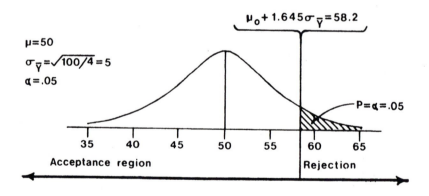

FIG. 4.4.   DIAGRAM FOR A TRUE NULL HYPOTHESIS

The shaded area represents the probability of a Type I error for this one-tailed test.

probabilities are seen to correspond to Z scores of 1.64 and 1.65, respectively. By linear interpolation, the value of Z corresponding to P = 0.4500 would be Z = 1.645. Since Z is the number of standard deviations distance from the mean, then the associated $\bar{Y}$ would be μ_o + 1.645 σ_Y or 50 + 1.645 (5) = 58.225. Therefore, *if* the H_0 is true and a sample mean is observed that is > 58.2, P will be less than α = 0.05 and the null hypothesis will be rejected, committing a Type I error.

Now suppose that H_0 is in fact false and that $\mu > \mu_o$, i.e., μ = 62.5. We would now diagram the situation as in Fig. 4.5. The shaded area now corresponds to the probability of committing a Type II error for the given situation. This area is the proportion of the *true distribution*, which lies in the acceptance region. The computation of the probability of a Type II error is possible only when the true mean and variance are known; therefore this procedure is useful only to help us understand hypothesis testing concepts. Again, consulting Appendix Table B.1 we can determine that β in this case is equal to 0.195 as shown in Table 4.2. However, if H_0 is false with $\mu > \mu_o$, i.e., μ = 52.5 (only a slight difference between μ and μ_o), the probability of committing a Type II error would be greatly increased (β = 0.874; Fig. 4.6).

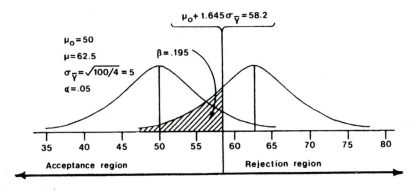

FIG. 4.5.   DIAGRAM OF TEST SHOWN IN FIG. 4.4 WHEN THE TRUE MEAN (μ = 62.5) DIFFERS FROM THE HYPOTHESIZED MEAN (μ = 50)

The shaded area represents the probability of a Type II error.

If one computes β for various differences between μ and μ_o, as in the last two examples, and plots $1 - \beta$ (power of the test) against $\mu - \mu_o$, one obtains what is known in statistics as the power function. The power function associated with the test diagrammed in Fig. 4.5 and 4.6 is presented in Fig. 4.7 for n = 4. Since sample size is the factor which is most likely to be controlled by the researcher, we have plotted power functions for n = 4 and n = 16 to give the reader a better understanding of the importance of the size of the sample.

TABLE 4.2.   COMPUTATION OF TYPE II ERROR (β) CORRESPONDING TO SHADED AREA OF FIG. 4.5.

Find the one-tailed critical value for $\alpha = 0.05$.

In Appendix Table B.1 find the Z value with 0.05 probability greater than Z.   Z = 1.645.

Given $\mu_o = 50$; $\sigma_{\bar{Y}} = 5$; and $Z = \dfrac{\bar{Y} - \mu_o}{\sigma_{\bar{Y}}}$. Then the critical

$$\bar{Y} = \mu_o + Z\,\sigma_{\bar{Y}}$$

$$= 50 + 1.645\,(5)$$

$$= 58.2$$

Compute the probability of a Type II error (β).

Given that $\mu = 62.5$, $\sigma_{\bar{Y}} = 5$, and the critical $\bar{Y} = 58.2$.

Then:

$$\beta = P\,(\bar{Y} < 58.2)$$

$$= P\left(Z < \frac{58.2 - 62.5}{5}\right)$$

$$= P\,(Z < -0.86)$$

from Appendix Table B.1

$$Z = 0.86 \Rightarrow 0.3051$$

$$\beta = 0.5000 - 0.3051 = 0.195$$

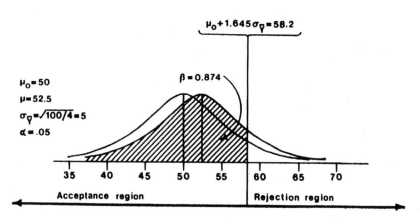

FIG. 4.6.   DIAGRAM OF THE FALSE NULL HYPOTHESIS ($\mu = 52.5$) FOR ONE-TAILED TEST

The shaded area represents the probability of a Type II error.

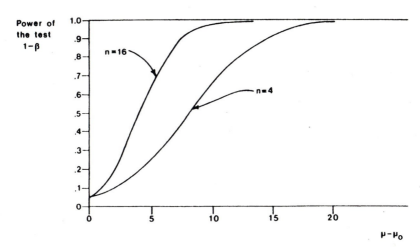

FIG. 4.7. POWER FUNCTION FOR SAMPLES OF n = 4 AND n = 16

In both cases $\mu = 50$, $\alpha = 0.05$, $\sigma^2 = 100$, and the alternate hypothesis is one-tailed.

Note that the minimum (when $\mu - \mu_o = 0$) is equal to α (0.05). Let's examine the effect of increasing sample size on hypothesis testing. For comparison assume that the true population mean (μ) is 5 units greater than the hypothesized mean (μ_o). Given the conditions of this test of hypothesis (Fig. 4.7), the probability of concluding that $\mu > \mu_o$ is ~ 0.25 if n = 4, but increases to ~ 0.65 if n = 16. Suppose that the researcher wanted to be very sure (say 0.95) that a difference of 5 units could be detected (H_o rejected if $\mu = 55$). In a similar manner power curves could be constructed for even larger sample sizes, until a sample size is found which has a power of the test of 0.95 when $\mu - \mu_o = 5$.

Four other factors in the hypothesis testing procedure influence the power of the test. These are the selection of a one- or two-tailed alternative, selection of the level of α, changes in sample size, and changes in variance. To understand the effect of each of these, let's use the conditions given for Fig. 4.5 as a reference and examine changes in each of these factors.

In Fig. 4.8 the researcher has elected to use a two-tailed test rather than the one-tailed test of Fig. 4.6. As in the one-tailed case, acceptance and rejection regions must be computed; however, in the two-tailed case the probability of α is equally assigned to the two tails creating two rejection regions. The probability of β is still found by computing the area of the true population in the acceptance region for the null hypothesis. Note that the selection of a two-tailed test resulted in an increase in β relative to the corresponding one-tailed test.

Figure 4.9 illustrates the effect of increasing α, for example, from 0.05 to 0.10. Notice the decrease in β compared to Fig. 4.6.

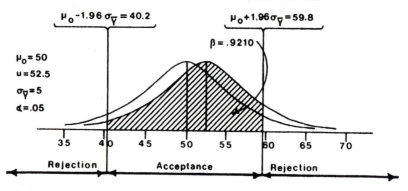

FIG. 4.8.   DIAGRAM OF THE FALSE NULL HYPOTHESIS (μ = 52.5) FOR A TWO-TAILED TEST

The shaded area represents the probability of a Type II error.

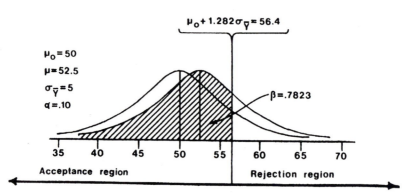

FIG. 4.9.   DIAGRAM OF THE FALSE NULL HYPOTHESIS (μ = 52.5) FOR A ONE-TAILED TEST

The shaded area represents the probability of a Type II error.

In some situations the researcher may be able to decrease variance (i.e., by use of an improved measurement technique). Figure 4.10 illustrates the effect of decreasing variance compared to Fig. 4.6. That is, decreasing variance decreases probability of committing a Type II error (β).

As shown in Fig. 4.7, the effect of increasing sample size is to decrease β. You may wish to diagram the effect of increasing sample size and compute the probability of β. For example, try n = 16, leaving all other parameters as given in Fig. 4.6. Compare your figure and the value of β given with Fig. 4.10 and note the similarity.

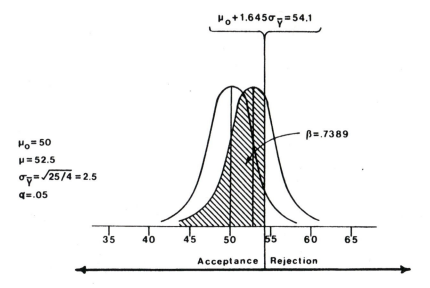

$$\mu_0 + 1.645\sigma_{\bar{Y}} = 54.1$$

$\beta = .7389$

$\mu_0 = 50$

$\mu = 52.5$

$\sigma_{\bar{Y}} = \sqrt{25/4} = 2.5$

$\alpha = .05$

35   40   45   50   55   60   65

Acceptance | Rejection

FIG. 4.10. DIAGRAM OF THE FALSE NULL HYPOTHESIS (μ = 52.5) FOR A ONE-TAILED TEST

The shaded area represents the probability of a Type II error.

Confidence Interval Estimation

A *point estimate* is a single number stated as an estimate of some quantitative characteristic of the population. Although the point estimate provides some information about the population, it is not very useful by itself since there is no indication of its reliability.

An interval estimate is a statement that a population parameter has a value lying between two specified limits. When an interval estimate is constructed in such a way that the probability of the two limits including the population parameter is known it is referred to as a *confidence interval estimate*.

Suppose that we would like to determine limits of the population mean μ such that we are $100(1 - \alpha)\%$ certain that the interval covers the true value of μ.

We know that:

$$Z = \frac{\bar{Y} - \mu}{\sigma/n^{1/2}}$$ is a normally distributed variable when sample size is large.

Thus we can write the following inequality expecting it to hold true for $100(1 - \alpha)\%$ of all possible samples drawn from the distribution.

$$P\left(-Z_{\alpha/2} \leq \frac{\bar{Y} - \mu}{\sigma/n^{1/2}} \leq Z_{\alpha/2} \right) = 1 - \alpha$$

Multiplying each term by $-\dfrac{\sigma}{n^{1/2}}$ and adding $\bar{Y}$ we obtain a confidence interval statement for μ.

$$P(\bar{Y} - Z_{\alpha/2}\,\sigma/n^{1/2} \leq \mu \leq \bar{Y} + Z_{\alpha/2}\sigma/n^{1/2}) = 1 - \alpha$$

or

$$P(\bar{Y} - Z_{\alpha/2}\,\sigma_Y \leq \mu \leq \bar{Y} + Z_{\alpha/2}\sigma_Y) = 1 - \alpha$$

The quantity $100(1 - \alpha)\%$ is called the confidence coefficient, $\bar{Y} - Z_{\alpha/2}\sigma_Y$ and $\bar{Y} + Z_{\alpha/2}\sigma_Y$ are the lower and upper confidence limits, denoted L_1 and L_2, respectively. The confidence interval length is $L_2 - L_1$. Since $\bar{Y}$ changes from sample to sample, the confidence interval is in fact a collection of intervals. The population parameter μ is a fixed quantity and is expected to be included in $100(1 - \alpha)\%$ of the interval estimates calculated from the above inequality.

Let's return to the two examples that were used to illustrate the hypothesis testing procedure. For the first example, concerned with the protein content of a cereal, the population variance was reported to be 2.25% and the sample mean on a random sample of 100 observations was $\bar{Y} = 11.8\%$. The 95% confidence limits for the population mean would be:

$$
\begin{aligned}
L_1 &= \bar{Y} - Z_{0.05/2}\sigma/n^{1/2} \\
&= 11.8 - 1.96\,(1.5/100^{1/2}) \\
&= 11.8 - 0.29 \\
&= 11.5 \\
L_2 &= 11.8 + 0.29 \\
&= 12.1
\end{aligned}
$$

Having computed the 95% confidence limits for true protein content, we can make the following statement: "We are 95% certain that the interval from 11.5 to 12.1% protein includes the true percentage protein in the new cereal grain." Two comments should be made about the confidence statement.

(1) The statement makes inference about the population parameter, not about other samples.
(2) The uncertainty in the confidence statement is a result of sampling.

In the second sample we wish to find an estimate of the numbers of eggs per plant on a producer's vegetable crop. The sample mean of 80 plants was 79.0 eggs/plant with a standard deviation of $\sigma = 40$. The 80% confidence limits are:

$$L_1 = \bar{Y} - Z_{\alpha/2}\sigma_{\bar{Y}}$$
$$= 79.0 - 1.28\,(4.47)$$
$$= 79.0 - 5.7$$
$$= 73.3$$
$$L_2 = 79.0 + 5.7$$
$$= 84.7$$

The concept of what happens with repeated sampling is a useful one for the student of statistics to understand. When we construct, for example, an 80% confidence interval, we should recognize that with repeated sampling from the population we would expect 20% of our confidence intervals to fail to contain μ while 80% of the intervals would be expected to contain μ. Figure 4.11 illustrates this repeated sampling concept. Each horizontal bar represents the 80% confidence interval for a single sample of size n. Note how the samples vary around the true μ and as expected sometimes fail to cover μ.

In the grain example, the confidence interval includes the hypothesized value; while in the vegetable crop, the hypothesized value lies outside the confidence limits. When confidence limits are calculated for the level of α used in the hypothesis testing procedure, the two procedures will agree. That is, when H_0 is accepted, the confidence interval will include the hypothesized value, and when H_0 is rejected, the confidence interval will not include the hypothesized value.

SMALL SAMPLE INFERENCE

The statistical techniques presented in the first part of this chapter deal with the following kinds of samples:

(1) Large samples (at least n = 30 and preferably n > 100) from any distribution. If variance is unknown, substitution of sample variance for population variance generally provides an adequate approximation.
(2) Samples of any size drawn from a normal distribution with known variance.

However, in most cases σ is unknown and sample size may be too small to allow the use of sample variance instead of population variance. We shall describe a new distribution known as the Student's t distribution which permits exact inferences to be made, when S^2 is computed from a sample of any size.

Student's t

The equation for Student's t is very similar to the equation for Z.

$$t = \frac{\bar{Y} - \mu}{S/n^{1/2}} = \frac{\bar{Y} - \mu}{S_{\bar{Y}}}$$

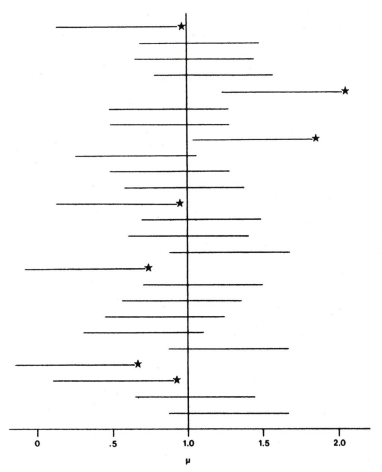

FIG. 4.11. TWENTY-FIVE 80% CONFIDENCE INTERVALS FOR THE MEAN,
DRAWN AT RANDOM FROM A NORMAL DISTRIBUTION WITH μ = 1, σ = 0.7,
AND n = 5

Intervals marked with a * failed to include the true population mean.

Student's t differs from Z in two very important respects:

(1) The t equation makes use of the sample standard deviation.
(2) The t distribution is not a single distribution, but rather there exists a separate distribution for every possible sample size from n = 2 to n = ∞.

As sample size approaches ∞, the t distribution approaches the standard normal distribution. However, with small sample size, the t distribution tends to be flatter, and takes on a wider range of values than the normal distribution. This is due to the fact that the sample standard deviation is located in the divisor of the t equation. Since S is a sample statistic, it too is subject to random variation and will at times overestimate, and at other times underestimate, σ. Thus, t itself is subject to greater variation than Z, particularly when sample size is small.

Student's t distributions are presented in Appendix Table C.1. To locate a t value in the table, two indices of t are required. These are α and the degrees of freedom (n − 1). When writing a tabular t value, these indices are generally written as subscripts to the $t_{\alpha(n-1)}$. Therefore, if we wish to locate the 5% t value for a sample of size 10, we would look for $t_{0.05(9)} = 2.262$, that is, the column marked 0.050 and row with degrees of freedom marked 9. Thus α values are represented by columns while the degrees of freedom correspond to rows.

In contrast to Appendix Table B.1 which reports the area under the curve from the mean (i.e., 0) to the desired value of Z, Appendix Table C.1 reports the critical value associated with a specific level of significance for the given degrees of freedom. In addition, it should be noted that the values reported in Appendix Table C.1 are for a two-tailed test.

In the limit (i.e., when the degrees of freedom = ∞), Student's t is normally distributed. This can be seen by comparing the entries in the last row of Appendix Table C.1 with the corresponding entries in Appendix Table B.1. For example, the value of Student's t for α = 0.05 and infinite degrees of freedom is 1.95996. In other words, for a two-tailed distribution and the critical values of ±1.95996, each tail will contain an area under the curve equal to 0.025 (i.e., α/2). In contrast, Appendix Table B.1 indicates that for a Z of 1.96, the area under the curve between the mean (i.e., 0) and the Z value (i.e., 1.96) is 0.4750.

Consequently, the tail will contain an area equal to 0.0250 (i.e., 0.5000 − 0.4750). Since Z is symmetric, ±1.96 will contain an area under the curve equal to 0.9500 (0.4750 + 0.4750), and the sum of two tails will be 0.0500, which corresponds to the α level for the middle column of Appendix Table C.1. The reader can make similar comparisons between Appendix Tables B.1 and C.1 by examining the other t values from the bottom row of Appendix Table C.1.

To utilize the t distribution for computations of confidence limits of μ we would write:

$$P(\bar{Y} - t_{\alpha(n-1)} S_Y \leq \mu \leq \bar{Y} + t_{\alpha(n-1)} S_Y) = 1 - \alpha$$

Suppose the randomly selected packages of Susan's Sophisticated Sausages when analyzed for water content yield the following data:

Sample	% H_2O
1	32
2	29
3	33
4	29
5	33
6	30
7	31
ΣY_i	217
ΣY_i^2	6745

$$\Sigma(Y_i - \bar{Y})^2 = \Sigma Y_i^2 - \frac{(\Sigma Y_i)^2}{n} = 6745 - \frac{(217)^2}{7} = 18.000$$

$$S^2 = \frac{\Sigma(Y_i - \bar{Y})^2}{n - 1} = \frac{18.00}{6} = 3.000$$

$$S = (S^2)^{1/2} = 3.000^{1/2} = 1.732$$

$$S_{\bar{Y}} = \frac{S}{n^{1/2}} = \frac{1.73}{7^{1/2}} = 0.655$$

$$\bar{Y} = \frac{\Sigma Y_i}{n} = 31.0$$

The 99% confidence limits are determined as follows:

$$
\begin{aligned}
L_1 &= \bar{Y} - t_{0.01(6)} S_{\bar{Y}} \\
&= 31.0 - 3.707 \ (0.655) \\
&= 31.0 - 2.43 \\
&= 28.6 \\
L_2 &= 31.0 + 2.43 \\
&= 33.4
\end{aligned}
$$

In order to test the hypothesis that the μ percentage water does not exce 30%, we proceed as follows:

$$
\begin{aligned}
H_o &: \mu = \mu_o = 30.0 \\
H_1 &: \mu > \mu_o = 30.0 \\
&\quad \alpha = 0.05
\end{aligned}
$$

$$t = \frac{\bar{Y} - \mu_o}{S_Y} = \frac{31.0 - 30.0}{0.655} = 1.52$$

Examination of Appendix Table C.1 indicates that the observed value (t = 1.52) lies between $t_{0.5(6)}$ = 0.718 and $t_{0.1(6)}$ = 1.943. Since the table presents two-tailed t values, we must divide α by 2 to obtain the one-tailed values. Thus for our one-tailed test the observed t of 1.52 has a probability (P) between 0.05 and 0.25, resulting in acceptance of H_o, since P · 0.05 as specified in the test of hypothesis.

Another way of approaching this test would have been to state that for our one-tailed test with α = 0.05, the critical value of t from Appendix Table C.1 is 1.943. That is, if the null hypothesis is true, calculated values of t greater than 1.943 will occur 5% of the time (remember that Appendix Table C.1 reports critical values for a two-tailed test, and we are dealing with a one-tailed test). Since our calculated value of t (i.e., 1.52) is less than the critical value reported in the table (i.e., 1.94), we accept the null hypothesis.

SUMMARY

The first four chapters of this text have attempted to lay a foundation of basic concepts essential to using and understanding statistical procedures. The remaining chapters build on these basic concepts and illustrate a variety of applications.

PROBLEMS

4.1. A researcher wishes to conduct a two-tailed test of hypotheses that the population mean (μ_o) equals 12. It is known that the population variance (σ^2) is 40. The researcher draws a random sample of 20 items from the population and calculates the sample mean ($\bar{Y}$).

   (a) State the null and alternative hypotheses.
   (b) Suppose the sample mean equals 9.1; compute the test statistic and determine its probability.
   (c) If the researcher chooses α = 0.05, should he accept or reject the null hypothesis? Write a brief statement of conclusion.

4.2. Given the same population parameters and sample statistics as in 4.1 (μ_o = 12, σ^2 = 40, n = 20, and $\bar{Y}$ = 9.1) how would each of the following have changed the test of hypotheses?

   (a) α = 0.01, rather than α = 0.05.
   (b) H_o: $\mu = \mu_o$
       H_1: $\mu > \mu_o$, rather than a two-tailed test.

4.3. Given that H_o: μ = 50; H_1: $\mu > 50$; α = 0.05; σ^2 = 100; and n = 16; compute β supposing that in fact μ = 62.5. Diagram the test of hypotheses.

4.4. The thickness of the shell of 10 randomly selected eggs was measured in mm. Test the hypothesis that the mean thickness is 1.5. The following measurements were recorded: 1.2; 1.6; 0.9; 1.3; 1.9; 1.4; 1.3 1.0; 1.7; 1.6.

4.5. Suppose that the concentration of a certain material cannot exceed 16 ppm in a given food product. Given the following data, test appropriate hypotheses at the α = 0.01 level of significance: 16.1; 15.9; 16.4 15.9; 16.1; 16.2; 16.0; 16.1; 15.8; 16.4; 16.3; 16.0.

REFERENCES

For a nonmathematical discussion see:

MENDENHALL, W. and OTT, L. 1976. Understanding Statistics, 2nd Edition. Duxbury Press, North Scituate, Mass. Chapters 7 and 8.

For general examples and discussion of hypothesis testing concepts, power of the test and "interval estimation" see:

LINDGREN, B.W. 1975. Basic Ideas of Statistics. Macmillan Publishing Co., New York. Chapters 6–8.

OSTLE, B. and MENSING, R.W. 1975. Statistics in Research, 3rd Edition Iowa State Univ. Press, Ames. Chapters 6 and 7.

SOKAL, R.R. and ROHLF, F.J. 1969. Biometry. W.H. Freeman and Co. San Francisco. Chapter 7.

For a more mathematical and theoretical approach see:

ANDERSON, R.L. and BANCROFT, T.A. 1952. Statistical Theory in Research. McGraw-Hill, New York. Chapters 10 and 11.

Part II

Statistics in Research

5

Comparisons Involving 2 Samples

In Chapter 4 we introduced the concept of hypothesis testing. To illustrate the hypothesis testing problem, and to study concepts related to Type I and Type II errors, we examined the test of hypotheses for a single sample mean. In research, it is rare that an experiment will involve a single sample. In this chapter we will concern ourselves with comparisons of means and variances from two samples. This situation is of considerably greater interest in research than the single sample tests. Data for the two-sample case may arise from either a survey involving two groups or from a designed experiment involving the comparison of two treatments. The use of the word "treatment" is in a general sense, in that one of the treatments could be a nontreated group or other appropriate control group. In Chapters 7 through 10 we shall concern ourselves with a technique with more general application known as the analysis of variance.

COMPARISON OF 2 SAMPLE MEANS

In this section we are concerned with the test of the null hypothesis, H_o: μ_1 = μ_2. The same null hypothesis may also be written as H_o: $\mu_1 - \mu_2 = \mu_D$, where μ_D = 0. This form of the null hypothesis illustrates some of the flexibility of the two sample tests of means, since the hypothesized difference (μ_D) may be any desired value, although it is commonly zero.

The alternative hypothesis may be one-tailed or two-tailed, just as in the single sample case. Also, as in the single sample case, the computation of the test statistic is the same regardless of the alternative. However, the determination of the probability of the test statistic does depend on whether the test is one- or two-tailed.

Two sample data may arise from either independent or paired samples. In the case of independent samples, observations included in the first sample are not associated with observations in the second sample. That is, observation one of sample one has no more relation to observation one of sample two than it has to any other observation in sample two. On the other hand, paired samples imply that each observation on sample one has a corresponding observation on sample two, which is similar in one or several respects, except for the treatment effect being studied. One of the strongest cases of pairing is the before and after experiment (called self-pairing). In a paired study, it is the difference between responses of paired subjects which is of interest to the researcher.

In the comparison of independent samples, it is assumed that the variances of the two sampled populations are equal ($\sigma_1^2 = \sigma_2^2$). The researcher should, in application, examine the validity of the assumption. If the assumption is not met, the researcher may resort to data transformation, distribution-free methods, or, as will be presented later, an approximate test statistic.

If the null hypothesis is $H_o: \mu_1 - \mu_2 = \mu_D$ (or $H_o: \mu_1 = \mu_2$), then the general form of the t-test for the two-sample case is as follows:

$$t = \frac{(\bar{Y}_1 - \bar{Y}_2) - (\mu_1 - \mu_2)}{S_{\bar{Y}_1 - \bar{Y}_2}} = \frac{\bar{D} - \mu_D}{S_{\bar{D}}}$$

This t statistic is the same as that presented in the previous chapters since $S_{\bar{Y}_1 - \bar{Y}_2} = S_{\bar{D}}$ is the standard error of the difference between the means, while $\bar{D}$ is the mean difference. As we shall see, the equation for $S_{\bar{D}}$ is different for paired and unpaired experiments, for equal or unequal variances, and can be simplified for samples with equal n. The following examples illustrate the application of the two-sample test of means.

Independent Samples with Equal Variances

Suppose that a researcher wished to test the hypothesis that added sucrose intake increases systolic blood pressure. When variances are equal, the common population σ^2 is estimated as:

$$S^2 = \frac{(n_1 - 1)S_1^2 + (n_2 - 1)S_2^2}{(n_1 - 1) + (n_2 - 1)}$$

where

$$S_1^2 = \frac{\Sigma(Y_1 - \bar{Y}_1)^2}{n_1 - 1} \quad \text{(definitional formula)}$$

$$= \frac{\Sigma Y^2_1 - \dfrac{(\Sigma Y_1)^2}{n_1}}{n_1 - 1} \quad \text{(computing formula)}$$

and

$$S_2^2 = \frac{\Sigma(Y_2 - \bar{Y}_2)^2}{n_2 - 1}$$

In this way, the information from the first sample (i.e., the Y_{1i}) and the second sample (i.e., the Y_{2i}) is pooled to provide a single estimate of the common variance.

It is a common convention to report observations with a capital letter (e.g., Y) and the deviation from the mean as a lower case letter (e.g., y). In other words, $y = Y - \bar{Y}$. Such a convention simplifies the presentation of formulas. An occasional lapse in this convention occurs when the mean is reported as $\bar{y}$ as well as $\bar{Y}$. Throughout this text, lower case letters (e.g., x, y) will represent deviations from the mean (i.e., $x = X - \bar{X}$ and $y = Y - \bar{Y}$). The mean will be represented with both upper and lower case letters (i.e., $\bar{x} = \bar{X}$ and $\bar{y} = \bar{Y}$).

Consequently, the formula just presented for the common variance (i.e., S^2) can be written as:

$$S^2 = \frac{(n_1 - 1)S_1^2 + (n_2 - 1)S_2^2}{(n_1 - 1) + (n_2 - 1)} = \frac{\Sigma y_1^2 + \Sigma y_2^2}{n_1 + n_2 - 2}$$

and $S_{\bar{D}}$

$$S_{\bar{D}} = \left[S^2 \left(\frac{1}{n_1} + \frac{1}{n_2} \right) \right]^{1/2} = \left[S^2 \left(\frac{n_1 + n_2}{n_1 n_2} \right) \right]^{1/2}$$

The data and calculations for the proposed problem are presented in Table 5.1. The sample evidence supports rejection of the null hypothesis ($P < \alpha$); therefore we conclude that sucrose intake did increase systolic blood pressure.

Although the researcher is interested in the increase in blood pressure, the null hypothesis must be stated in a manner that will permit the calculation of the test statistics. In this case, as in most cases, the null hypothesis states that there is no difference between the means. That is, $H_o: \mu_2 - \mu_1 = 0$ is the same as $H_o: \mu_2 = \mu_1$.

Since the researcher is interested only in the possibility of increased blood pressure due to treatment, he proposes a one-tailed test. It is customary, when comparing a treatment to a control, to state the test in terms of the treatment mean minus the control mean. Therefore, $H_1: \mu_2 - \mu_1 > 0$ which is equivalent to saying $H_1: \mu_2 > \mu_1$. However, we could just as correctly state the alternative hypothesis as $H_1: \mu_1 - \mu_2 < 0$, or $H_1: \mu_1 < \mu_2$, with the test statistic stated as

$$t = \frac{(\bar{Y}_1 - \bar{Y}_2) - \mu_D}{S_{\bar{D}}},$$ with large negative t values leading to rejection of the null hypothesis.

The researcher might also wish to calculate confidence intervals for the treatment means as in Chapter 4, or for the difference between control and treatment means. For the 90% confidence limits of the difference between two means he computes:

TABLE 5.1.   EXAMPLE OF TWO SAMPLE T-TEST ($\sigma_1^2 = \sigma_2^2$ AND $n_1 \neq n_2$)

	Systolic Blood Pressure After 13 Weeks of Treatment	
	Controls	Sucrose (50 g/Day
	106	110
	98	134
	108	122
	104	104
	120	118
	124	131
	108	114
	96	
	100	
ΣY	964	833
ΣY^2	103,976	99,837
$\Sigma y^2 = \Sigma Y^2 - (\Sigma Y)^2/n$	720.9	710.0
S^2	90.1	118.3
$df = n_1 + n_2 - 2$	8	6
$\bar{Y}$	107.1	119.0
S_Y	3.16	4.11

Hypotheses (one-tailed test):

H_o: $\mu_2 = \mu_1$
H_1: $\mu_2 > \mu_1$

Pooled variance:

$$S^2 = \frac{\Sigma y_1^2 + \Sigma y_2^2}{n_1 + n_2 - 2} = \frac{720.9 + 710.0}{9 + 7 - 2} = 102.2$$

Standard error of the difference:

$$S_{\bar{D}} = \left[S^2 \left(\frac{n_1 + n_2}{n_1 \cdot n_2} \right) \right]^{1/2} = \left[102.2 \left(\frac{9 + 7}{9 \cdot 7} \right) \right]^{1/2} = 5.09$$

Test statistic:

$$t = \frac{(\bar{Y}_2 - \bar{Y}_1) - \mu_D}{S_{\bar{D}}} = \frac{(119.0 - 107.1) - 0}{5.09} = 2.34$$

Probability of test statistic (df = 14):

From Appendix Table C.1, we find the tabular value associated with an α equal to 0.05 Since Appendix Table C.1 is calculated for a two-tailed test, we must use the column headed 0.1. The tabular value of t for 14 degrees of freedom and this level of probability is 1.761. Since our calculated value of t (2.34) is greater than the critical value listed in Appendix Table C.1, we reject the null hypothesis.

$$L_1 = (\bar{Y}_2 - \bar{Y}_1) - t_{\alpha(df)} S_D$$
$$= +11.9 - 1.761 (5.09) = 2.9$$
$$L_2 = +11.9 + 1.761 (5.09) = 20.9$$

As expected, by the rejection of H_o, the confidence limits do not include zero. The researcher would likely choose the 90% confidence limits since they represent 5% in each tail and their result would correspond to the 5% one-tailed test of hypothesis.

If the sample sizes are equal, that is $n_1 = n_2 = n$, then S_D can be shown to be:

$$S_{\bar{D}} = \left[S^2 \left(\frac{1}{n_1} + \frac{1}{n_2} \right) \right]^{1/2} = \left[\frac{2S^2}{n} \right]^{1/2}, \text{with } 2(n-1)\text{df}.$$

An example is presented in Table 5.2.

TABLE 5.2.   EXAMPLE OF TWO SAMPLE T-TEST ($\sigma_1^2 = \sigma_2^2$ AND $n_1 = n_2$)

	Serum Cholesterol (mg %)	
	Non-Contraceptive	Oral Contraceptive
	163	132
	157	147
	245	177
	169	126
	93	213
	110	219
ΣY	937	1,014
ΣY^2	160,553	179,568
$\Sigma y^2 = \Sigma Y^2 - (\Sigma Y)^2/n$	14,225	8,202
S^2	2,845	1,640
$\text{df} = 2(n - 1)$	5	5
$\bar{Y}$	156	169
$S_{\bar{Y}}$	21.8	16.5

Hypotheses (two-tailed test):

H_o: $\mu_1 = \mu_2$
H_1: $\mu_1 \neq \mu_2$

Pooled variance:

$$S^2 = \frac{\Sigma y_1^2 + \Sigma y_2^2}{n_1 + n_2 - 2} = \frac{14,225 + 8202}{6 + 6 - 2} = 2242.7$$

Standard error of the difference:

$$S_{\bar{D}} = \left(\frac{2S^2}{n} \right)^{1/2} = \left[\frac{2(2242.7)}{6} \right]^{1/2} = 27.3$$

Test statistic:

$$t = \frac{(\bar{Y}_1 - \bar{Y}_2) - \mu_D}{S_{\bar{D}}} = \frac{(156 - 169) - 0}{33.5} = -0.476$$

Probability of test statistic (df = 10):

This time we select an α equal to 0.05, but for a two-tailed test. That is, we examine the column headed 0.05 and the row for 10 degrees of freedom. The critical value is 2.228. If the calculated value of t (i.e., -0.476) is greater than 2.228 or less than -2.228, we reject the null hypothesis.

Since the calculated value of t lies between these critical values, we accept the null hypothesis.

Independent Samples with Unequal Variance

If the assumption of equal variances is found to be untenable, an exact procedure is unavailable. A test of equality of variances is given later in this chapter. The test of homogeneity of variances should be computed before using the t test on independent samples. The following approximation is satisfactory for most cases. Since the variances are not equal, a pooled variance is not appropriate. The test statistic is:

$$t' = \frac{(\bar{Y}_1 - \bar{Y}_2) - \mu_D}{\left(\dfrac{S_1^2}{n_1} + \dfrac{S_2^2}{n_2}\right)^{1/2}}, \text{ with}$$

$$df' = \frac{\dfrac{S_1^2}{n_1} + \dfrac{S_2^2}{n_2}}{\dfrac{(S_1^2/n_1)^2}{n_1 - 1} + \dfrac{(S_2^2/n_2)^2}{n_2 - 1}}$$

The computed df' is generally not an integer and since the test is not an exact test, we generally use the next lower integer value to determine the probability of t' based on the Student's t distribution. The data of Table 5.3 are used to illustrate this application.

Paired Samples

For paired samples, the computation procedure is concerned only with the difference between pairs. Once these differences are computed, the test procedure is not unlike the single sample test of hypothesis. The only assumptions are that the pairs are independent and that the differences are from a normally distributed population. Even this assumption is not generally of great concern if we can use our knowledge of the central limit theorem. In addition, experience has shown that considerable departures from normality do not seriously affect the sample t distribution.

The t statistic for n paired samples is:

$$t = \frac{\bar{D} - \mu_D}{S_{\bar{D}}}, \text{ with } n - 1 \text{ df.}$$

The variance of the D's is:

$$S_D^2 = \frac{\sum\limits_{i=1}^{n} (D_i - \bar{D})^2}{n - 1}, \text{ where } \Sigma(D_i - \bar{D})^2 = \Sigma d_i^2 = \Sigma D_i^2 - \frac{(\Sigma D_i)^2}{n}$$

TABLE 5.3.   EXAMPLE OF TWO SAMPLE T-TEST $(\sigma_1^2 \ne \sigma_2^2)$

| | Glycogen Content (mg/g of Tissue) of Chicken Breast | |
	Uncooked	Microwave Cooked
	28	12
	17	7
	36	11
	23	10
	27	11
$\Sigma Y =$	131	51
$\Sigma Y^2 =$	3627	535
$\Sigma y^2 = \Sigma Y^2 - (\Sigma Y)^2/n$	194.8	14.8
S^2	48.7	3.7
$\bar{Y}$	26.2	10.2
$S^2_{\bar{Y}} = S^2/n$	9.74	0.74
$S_{\bar{Y}}$	3.12	0.86

Hypotheses (two-tailed test):

H_o: $\mu_1 = \mu_2$
H_1: $\mu_1 \ne \mu_2$

Test statistic:

$$t' = \frac{(\bar{Y}_1 - \bar{Y}_2) - \mu_D}{\left(\dfrac{S_1^2}{n_1} + \dfrac{S_2^2}{n_2} \right)^{1/2}} = \frac{(26.2 - 10.2) - 0}{(9.47 + 0.74)^{1/2}} = 5.01$$

$$df' = \frac{(S^2_{\bar{Y}_1} + S^2_{\bar{Y}_2})^2}{\dfrac{(S^2_{\bar{Y}_1})^2}{n_1 - 1} + \dfrac{(S^2_{\bar{Y}_2})^2}{n_2 - 1}} = \frac{(9.74 + 0.74)^2}{\dfrac{(9.74)^2}{4} + \dfrac{(0.74)^2}{4}} = 4.6 \text{ or } 4$$

Probability of test statistic (df = 4).

With $\alpha = 0.05$ and 4 degrees of freedom, the critical values from Appendix Table C.1 for this two-tailed test are ± 2.776.
Since the calculated value of t (5.01) is well outside the critical values (± 2.776), we reject the null hypothesis.

$$S_{\bar{D}} = \left(\frac{S_D^2}{n} \right)^{1/2}$$

An example of the test of hypothesis for paired samples is given in Table 5.4. It can be shown that $S_D^2 = S_1^2 + S_2^2 - 2\text{Cov}(Y_1, Y_2)$. The covariance of two variables $[\text{Cov}(Y_1, Y_2)]$ is a measure of the association between the paired observations. If the pairing is successful, large values of Y_1 will be associated with large values of Y_2 and small values of Y_1 with small values of Y_2. Thus, the covariance will be a large positive value. For independent samples with equal variances $S_D^2 = 2S^2$; however, in the successfully paired experiment $S_D^2 < 2S^2$, thus increasing sensitivity.

TABLE 5.4.   EXAMPLE OF A PAIRED T-TEST

Groundhog	Temperature (°C) of Supraspinatus Muscle of Groundhogs at Initiation of Caudal Vasodilation Spontaneous Arousal	Evoked Arousal	Difference
1	36.2	34.2	+2.0
2	34.4	33.2	+1.2
3	35.0	34.6	+0.4
4	35.6	36.0	-0.4
5	38.0	36.2	+1.8
6	34.2	32.8	+1.4
7	36.6	35.0	+1.6
8	33.2	33.4	-0.2
$\bar{Y}$	35.40	34.42	+0.98
Σd^2			5.955
S_D^2			0.851
$S_{\bar{D}}$			0.33

Hypotheses (two-tailed):

H_o: $\mu_1 = \mu_2$
H_1: $\mu_1 \neq \mu_2$

Test statistic:

$$t = \frac{\bar{D}}{S_{\bar{D}}} = \frac{+0.98}{0.33} = +2.97$$

Probability of test statistic (df = 7).

With $\alpha = 0.05$ and 7 degrees of freedom, the critical values for this two-tailed test from Appendix Table C.1 are ± 2.365. Therefore, we reject the null hypothesis of no difference.

TEST OF 2 SAMPLE PROPORTIONS

Another distribution commonly observed in food science and agriculture is that of the binomial variable. When each independent observation can be classified into one of only two possible classes, the resulting variable is said to follow the binomial distribution. For example, consider the following possible outcomes: fertile vs. sterile, life vs. death, response vs. no response, resistant vs. susceptible, success vs. failure, etc. These kinds of variables are frequently examined as a proportion of the total number of observations in each of the classes.

The procedure we shall use is referred to as the normal approximation to the binomial. We can again make use of the central limit theorem to approach normality since the proportion is in fact the mean number in the class per observation. The size of sample necessary to assume normality is dependent on values of p and q. When values of p and q lie close to 0.5, relatively few observations are required to approximate the normal distribution as compared to when the values of p or q lie close to 0 or 1. A commonly recommended rule is that if np and nq are greater than 10, the distribution of the proportions is approximately normal.

The point estimate of the binomial parameter p is computed as as $\hat{p} = \dfrac{\text{number of successes}}{\text{number of trials}} = \dfrac{Y}{n}$. The "hat" over the p denotes the sample estimate of the population parameter. It can be shown that if $\hat{p}$ is the proportion in one class and $\hat{q}$ is the proportion in the second class, that the variance of $\hat{p}$ is $\hat{p}\hat{q}$ and that the standard error ($S_{\hat{p}}$) equals $(\hat{p}\hat{q}/n)^{1/2}$

For the two-sample comparison, the test statistic for the normal approximation is $Z = \dfrac{\hat{p}_1 - \hat{p}_2}{\sigma_{\hat{p}_1 - \hat{p}_2}}$. The standard error of the difference between two binomial proportions is $\sigma_{\hat{p}_1 - \hat{p}_2} = \left[\hat{p}\hat{q}\left(\dfrac{1}{n_1} + \dfrac{1}{n_2} \right) \right]^{1/2}$ where $\hat{p} = \dfrac{Y_1 + Y_2}{n_1 + n_2}$ and $\hat{q} = 1 - \hat{p}$. An example of the two-sample test of hypothesis by normal approximation is presented in Table 5.5.

TESTING THE EQUALITY OF VARIANCES

In testing the hypothesis $\mu_1 = \mu_2$ for independent samples, the appropriate analysis depends on whether the group variances (σ^2) are equal or unequal. It should be apparent to most readers that some test is needed so that the researcher will know when to use the approximate t-test (t').

Suppose we wish to test the null hypothesis that $\sigma_1^2 = \sigma_2^2$. A new test statistic known as the F-test is required. The F-distribution is the ratio of the variances of two independent samples drawn from populations with a common variance. If the null hypothesis were true, we would expect the ratio of the sample variances to be close to one. How close to one the ratio should be is dependent on the degrees of freedom of the sample variances. The values tabulated in Appendix Tables D.1 and E.1 are for the 5% and 1% F-distributions, respectively. The location of an F value requires three indices, α, the numerator df (v_1), and the denominator df (v_2). The tabulated values represent the right half of the distribution.

In order to illustrate the use of the F-test for testing for homogeneity of variances, we will use the data presented in Table 5.3. In that example, we assume that $\sigma_1^2 \neq \sigma_2^2$. However, the test shown in Table 5.6 was conducted in order to reach that conclusion.

In Table 5.3, two independent samples of size 5 were used to estimate the population variance. The first estimate was $S_1^2 = 48.7$ with 4 degrees of freedom. The second estimate was $S_2^2 = 3.7$ with 4 degrees of freedom. The calculated value of F is the ratio of these independent estimates of σ^2.

The values for the F statistic in Appendix Tables D.1 and E.1 are similar to those in Appendix Table C.1 in that they report the critical value for a predetermined probability level. This concept can be seen in Fig. 5.1. As shown in Fig. 5.1, F can range between 0 and ∞. Our procedure is to select a predetermined probability for the Type I error and to link up the appropriate value from Appendix Table D.1 (or E.1) given our degrees of freedom. In this case, our degrees of freedom are $v_1 = 4$ and $v_2 = 4$. From Appendix Table

TABLE 5.5.   EXAMPLE OF TWO-SAMPLE Z TEST FOR BINOMIAL SAMPLES (NORMAL APPROXIMATION)

Response	Number of Animals	
	Control	Treated
Ovulation (Y_i)	14	20
No ovulation	18	10
Total (n_i)	32	32
$\hat{p}_i = Y_i/n_i$	0.438	0.625
$S_{\hat{p}} = (\hat{p}_i\hat{q}_i/n_i)^{1/2}$	0.088	0.086

Hypotheses (one-tailed test):

H_o: $p_T = p_C$
H_1: $p_T > p_C$

Standard error of the difference:

$$\sigma_{\hat{p}_T - \hat{p}_C} = \left[\hat{p}\hat{q}\left(\frac{1}{n_1} + \frac{1}{n_2}\right)\right]^{1/2}$$

$$= \left[(0.531)(0.469)\left(\frac{1}{32} + \frac{1}{32}\right)\right]^{1/2} = 0.125, \text{where}$$

$$\hat{p} = \frac{Y_1 + Y_2}{n_1 + n_2} = 0.531 \text{ and } \hat{q} = 1 - \hat{p} = 0.469$$

Test statistic:

$$Z = \frac{\hat{p}_T - \hat{p}_C}{\sigma_{\hat{p}_T - \hat{p}_C}} = \frac{0.625 - 0.438}{0.125} = 1.50$$

Probability of the test statistic (Appendix Table B.1):

Because we are dealing with the normal approximation of the binomial, we use the standard normal deviate (i.e., Z) to construct our test and evaluate its probability. We do this under the assumption that we know the true population standard deviation. Consequently, we use Appendix Table B.1 and do not concern ourselves with the question of degrees of freedom.

For $\alpha = 0.05$ with a one-tailed test, the critical value of Z is 1.645. This value is found by interpolation. We are interested in a tail probability of 0.05. This is equivalent to an area from 0 to Z of 0.4500. There is no entry in Appendix Table B.1 exactly equal to 0.4500. However, for $Z = 1.64$, the area from 0 to Z is 0.4495. For $Z = 1.65$, the area from 0 to Z is 0.4505. Through linear interpolation, we conclude that for $Z = 1.645$, the area from 0 to Z is 0.4500. Consequently, calculated values of Z greater than 1.645 would be expected only 5% of the time.

Our calculated value of Z is 1.50. Therefore, we would accept the null hypothesis of no difference.

D.1, we find that the critical value for F is 6.39. That is, if the null hypothesis is true, values greater than 6.39 will occur 5% of the time. In addition, small values of F would also indicate rejection of the null hypothesis for a two-tailed test of hypothesis. In this case a calculated value of 13.16 was obtained.

Even though it is not necessary in this case to determine the critical value of the left-hand tail (for small F values), we must remember to sum the

TABLE 5.6.   F-TEST FOR EQUALITY OF VARIANCE FOR THE PROBLEM OF TABLE 5.3

Hypotheses:

$$H_o: \quad \sigma_1^2 = \sigma_2^2$$
$$H_1: \quad \sigma_1^2 \neq \sigma_2^2$$

Test statistic:

$$F = \frac{S_1^2}{S_2^2} = \frac{48.7}{3.7} = 13.16$$

Probability of test statistic ($v_1 = 4; v_2 = 4$).

From Appendix Table D.1 with $\alpha = 0.05$ and 4,4 degrees of freedom, the tabular value reported for F is 6.39. Since the test is two-tailed we must double the one-tailed probability or $\alpha = 0.10$. Since the calculated value of F (13.16) is greater than the tabular value of F (6.39), we would reject the null hypothesis and conclude that the variances are not equal at the 10% level of significance.

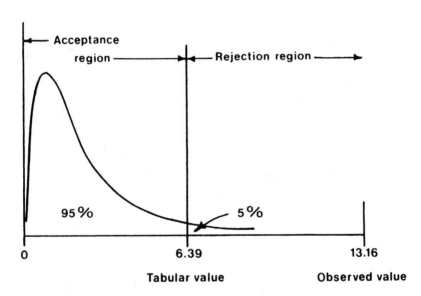

FIG. 5.1.   GRAPHIC PRESENTATION OF THE F-TEST

probabilities in the two-tails to determine the correct probability. The probability of a Type I error would equal 2(0.05) = 0.10. Consequently, we would reject the null hypothesis if the test of hypothesis were being tested at the 10% level of significance.

Many researchers use the numerically larger estimate of σ^2 as the numerator and the smaller as the denominator. The reason is that most F tables report only critical values associated with the right tail as illustrated in Fig. 5.1. However, if the left-hand tail is needed, it can be calculated as follows:

$$F_{1-\alpha(v_1, v_2)} = \frac{1}{F_{\alpha(v_2, v_1)}}$$

This calculation is illustrated in the next example.

For the problem summarized in Table 5.1, we assumed $\sigma_1^2 = \sigma_2^2$. However, the test shown in Table 5.7 was conducted in order to reach that conclusion.

In Table 5.1, two independent samples were used to estimate the (assumed) common variance. The first sample (i.e., the control) with 9 observations estimated $S_1^2 = 90.1$ with 8 degrees of freedom. The second sample (i.e., the treatment) with 7 observations estimated $S_2^2 = 118.3$ with 6 degrees of freedom. The question before us is whether or not these values could have been observed if the two populations possess a common variance.

TABLE 5.7.   EXAMPLES OF F-TEST FOR EQUALITY OF VARIANCE FOR THE PROBLEM OF TABLE 5.1

Test for data of Table 5.1.

Hypotheses:

H_o: $\sigma_1^2 = \sigma_2^2$
H_1: $\sigma_1^2 \neq \sigma_2^2$

Test statistic:

$$F = \frac{S_1^2}{S_2^2} = \frac{90.1}{118.3} = 0.762$$

Probability of test statistic ($v_1 = 8$; $v_2 = 6$).

Since S_1^2 is less than S_2^2, we need to calculate the left-hand tail of the F distribution.

$$F_{1-\alpha(v_1, v_2)} = \frac{1}{F_{\alpha(v_2, v_1)}}$$

For our case:

$$F_{0.95(8,6)} = \frac{1}{F_{0.05(6,8)}}$$

From Appendix Table D.1, $F_{0.05(6,8)} = 3.58$

$$F_{0.95(8,6)} = \frac{1}{3.58} = 0.279$$

In this case, calculated values which are smaller than our transformed tabular value indicate significance. Since our calculated value (0.762) is not smaller than our derived tabular value (0.279), we accept the null hypothesis and conclude that there is no difference between the two estimates of σ^2. Consequently, we are justified in pooling the data in order to derive a single estimate of the common variance as was done in Table 5.1.

Our hypotheses are:

$$H_o: \sigma_1^2 = \sigma_2^2$$
$$H_1: \sigma_1^2 \neq \sigma_2^2$$

Since the F tables report only large values of F (i.e., for the right-hand tail), we could arbitrarily place the larger estimate of σ^2 in the numerator. Consequently, our F statistics would be

$$F = \frac{S_2^2}{S_1^2} = \frac{118.3}{90.1} = 1.313$$

with 6 and 8 degrees of freedom.

From Appendix Table D.1 the tabular value for $F_{0.05(6,8)}$ is 3.58. With a calculated value of F (1.313) which is smaller than the critical value reported in the table (3.58), we would accept the null hypothesis of no difference. Therefore, we would be justified in pooling the data to develop a single estimate of the common variance as was done in Table 5.1.

It should be noted that the F-test always implies a two-tailed test since both the numerator and the denominator are squared terms. This is true in spite of the fact that only one tail of the distribution is used.

Although it is usually easier to simply place the larger estimate of σ^2 in the numerator and use the F tables as normally printed, it is possible to calculate the probability for the left-hand tail as shown in Table 5.7. Such a transformation will not change the outcome of the test because we are simply dealing with the reciprocals of the values used in the text. In addition, the added calculations introduce additional opportunities for errors.

SUMMARY

This chapter has examined several simple problems dealing with two sample comparisons. In addition, the F-test was introduced. The next chapter presents a number of basic principles concerning the statistical design of experiments. Chapters 7 through 10 present the procedure known as the analysis of variance, which is a generalized test for comparisons and utilizes the F-statistic.

PROBLEMS

5.1. Rainfall infiltration rates were observed for plots which had been planted continuously to soybeans or corn for several years. The researcher wishes to determine if the infiltration rates differ for the two cropping systems.

Continuous Soybeans	Continuous Corn
1.16	0.84
0.87	1.96
0.93	0.73
0.88	1.42
1.23	1.41
0.40	0.89
1.55	0.73

5.2. The mean body lengths of adult *Tropocyclops prasinus* were recorded at different sampling stations in two freshwater ponds. Is there sufficient evidence to indicate a difference in mean body length between the two ponds?

Jeffries Hills Pond	Milk Producers Pond
1.68	2.01
1.72	1.99
1.70	2.11
1.51	1.85
1.81	2.00
1.91	1.73
1.49	1.97
1.68	
1.54	
1.83	

5.3. Two populations of the insect *Epilachna varivestis* were observed for the soybean leaf consumption (cm^2). Test the hypothesis of no difference in mean consumption between the two populations.

Population I	Population II
327	40
265	125
65	233
441	148
236	103
198	79
242	188
155	97

5.4. A researcher measured the concentration of fecal coliforms (per 100 ml) in surface runoff from tall fescue-covered plots following 5 rainfall events. Plots were designated either control (nontreated) or sludge-treated. Is there evidence that fecal coliform concentration is greater on sludge-treated plots?

Rainfall	Control	Sludge
1	1.51	1.32
2	3.17	4.14
3	3.97	4.57
4	3.11	4.64
5	0.66	0.24

5.5. Obtaining seed germination of woody ornamentals is a major problem for nurserymen. Three hundred seeds of a certain species known for its hard seed coat were treated for one-half hour with concentrated sulfuric acid. Half of the seeds were then treated with 500 ppm of potassium gibberellate while the other half were left untreated. Conduct a test of hypothesis to determine if the potassium gibberellate treatment increased seed germination.

	Treatment	
Germination	½ hr H_2SO_4	½ hr H_2SO_4 + 500 ppm Potassium Gibberellate
Number successfuls	81	113
Number failures	69	37

REFERENCES

For additional examples, applications, and other two-sample techniques:

HUNTSBERGER, D.V. 1967. Elements of Statistical Inference, 2nd Edition. Allyn and Bacon, Boston. Chapters 7 and 8.

MENDENHALL, W. 1975. Introduction to Probability and Statistics, 4th Edition. Duxbury Press, N. Scituate, Mass. Chapter 7.

SNEDECOR, G.W. and COCHRAN, W.G. 1967. Statistical Methods, 6th Edition. Iowa State Univ. Press, Ames. Chapter 4.

STEEL, R.G.D. and TORRIE, J.H. 1960. Principles and Procedures of Statistics. McGraw-Hill Book Company, New York. Chapter 5.

Examples of t-test applications in recent literature:

CALKINS, C.R., SAVELL, J.W., SMITH, G.C. and MURPHY, C.E. 1980. Quality-indicating characteristics of beef as affected by electrical stimulation and postmortem chilling time. J. Food Sci. *45*, 1330.

ESBENSHADE, K.L. and DAY, B.N. 1980. The response of gilts reared in confinement to erogenous gonadotropin and estradiol benzoate. J. Anim. Sci. *51*, 668.

PONCET, C. and RAYSSIGUIER, Y. 1980. Effect of lactose supplement on digestion of lucerne hay by sheep. J. Anim. Sci. *51*, 180.

SAVELL, J.W., SMITH, G.C. and HUFFMAN, D.L. 1980. Cutting yields and palatability traits of hand-cut or frozen-tempered beef sub primals. J. Food Sci. *45*, 107.

SEGERSON, E.C. and JOHNSON, B.H. 1980. Selenium and reproductive function in yearling angus bulls. J. Anim. Sci. *51*, 395.

6

Concepts of Experimental Design

The concepts of experimental design considered in this chapter are those primarily concerned with comparative experiments, that is, experiments in which the researcher wishes to compare a number (two or more) of treatments or treatment combinations, which may include a control(s), or standard treatment. The researcher is not necessarily interested in the actual treatment means, but rather in the difference between treatment means.

SOME DEFINITIONS

Experimental Design

The experimental design is the *complete* sequence of steps involving randomization, assignment and application of treatments, collection of data, etc., for the experimental material, such that valid inferences may be drawn.

It is at this stage that the consulting statistician can make a meaningful contribution to one's research effort. If he is consulted only after completion of the study, regarding analysis of it, his options are generally limited or perhaps none at all.

In the design stage the consulting statistician should share the responsibility for making sure that objective inferences can be drawn, that valid estimates of experimental error can be computed, and that differences between treatments are in fact unbiased. Merely meeting the preceding definition of an experimental design is not necessarily sufficient. Since many experimental designs may lead to valid inferences, the statistician must also share the responsibility of identifying alternative designs which will fit the experimental limitations and provide the most (or the required) information with the least quantity of resources (cost). It is in these areas that the consulting statistician's knowledge and experience can best be used in one's research.

Experimental Unit

The experimental unit is the smallest unit of experimental material to which a treatment is assigned by a single act of randomization.

It is important that you think in terms of the treatment's being assigned to the experimental unit, and not the unit's being assigned to the treatment. An illustration of the problem would be helpful at this point.

Suppose that a researcher wishes to conduct an experiment involving 3 dietary treatments on beef cattle. Three pens are available, and each will accommodate 6 animals of the desired weight. The researcher proceeds to assign 6 animals to each treatment and a treatment group to each pen. The researcher believes that each animal is an experimental unit, and that there are 6 replicates. In fact, the treatment was assigned to a pen of animals, thus the pen (since all the animals in the pen receive the same treatment) constitutes the experimental unit. As a result, the experiment has no replication and cannot be analyzed without making certain assumptions which may be untenable.

Sampling Unit

The unit of experimental material on which the observation(s) is recorded is referred to as the sampling unit. The sampling unit may be the same as the experimental unit; the entire experimental unit may be divided into sampling units; or one or a few (less than the total) sampling units may be randomly sampled from the experimental unit. The sampling unit is sometimes also referred to as the observational unit.

Error

The use of the word error is, at the least, an unfortunate choice of terminology. However, it is in such common usage that about all that we can reasonably do is to define what we mean by the term error. First, let us say that it is not intended to imply that something is incorrect or wrong. The word error is simply another word for variation. We commonly use the word to refer to experimental error (variation) and sampling error (variation).

Experimental error is a measure of the variation between experimental units treated alike. This variation is due to both the inherent variability of the experimental material and failure of the experimental units to be handled (processed, measured, analyzed, etc.) identically, in spite of our best efforts. It is the presence of experimental error which has made statistics an integral part of research. In the absence of experimental error, statistical techniques would not be necessary since a single observation on each treatment would be sufficient to determine if treatment differences exist.

Experimental error may in some experiments be directly computable, while in other designs only an indirect measurement will be possible. In those analyses where experimental error has been measured directly, it is commonly termed the "within" variation. The term "within" refers to the variation between experimental units "within" treatments. When no direct calculation is possible, the experimental error is estimated by computing all assignable sources of variation and assuming that any variation left over represents experimental error. For these analyses, the experimental error is frequently termed "residual" or "remainder."

Sampling error is the variation between sampling units from the same experimental units. In many experiments, only a single sampling unit is used; or observations for sampling units from a single experimental unit are averaged to obtain a single value for each experimental unit. When only one observation per experimental unit is available, it does not mean that sampling variation does not exist, but rather that it simply cannot be measured. In most cases, the failure to estimate sampling variance does not affect our ability to draw valid inferences about differences between treatments.

In the cattle feeding experiment, it was noted that a pen of animals represents an experimental unit. If an observation is weight gain, and animals are weighed individually, then the animal is the sampling unit.

Treatment

We have used the word treatment without any real definition of the term. The treatment refers to a set of experimental conditions which will be applied (or associated with) the experimental units. One particular set may be, for example, a "control" set. The particular set used varies in a certain way from treatment to treatment, and it is the effect of these varying conditions in which we are interested.

A certain condition, when allowed to take on a number of possible levels, is called a factor. When an experiment contains more than one factor, each assigned more than one level, the treatment combinations form a factorial experiment. Factorial experiments are probably the most important sets of treatments in biological research, and Chapter 9 is devoted to their application, analysis, and interpretation.

PRINCIPLES OF EXPERIMENTAL DESIGN

Replication, randomization, and local control are generally referred to as the 3 basic principles of experimental design. To these we can also add 3 other concepts which, if they are not basic, are closely associated with the 3 basic principles, that is, sensitivity, orthogonality, and confounding. These 6 are so important to our understanding of experimental design that a section will be devoted to each one, even though important interrelationships exist among them.

Replication

A treatment is said to be replicated when it is assigned to more than one experimental unit in an experiment. In experimental design, a replicate is defined as one complete set of treatments (or treatment combinations) or repetition of the basic experiment. Note that replication requires multiple experimental units per treatment. It is not sufficient to have repeated readings on the experimental unit as was illustrated in the cattle feeding example. Even though weights on individual animals would provide 6 observations per experimental unit, the experiment is unreplicated. Replication concepts may be used in 3 ways in experimental design.

(1) Replication permits the estimation of experimental error, which is needed to evaluate the significance of the differences between the treatment means.
(2) Increasing numbers of replications increase the sensitivity by reducing the standard error of the difference between treatment means.
(3) Replication can be used to increase the scope of influence by incorporating a greater diversity of experimental material if appropriate local control is used to control experimental error.

To compute the correct probability for a test statistic a valid estimate of the experimental error must be available. As discussed in the cattle feeding example, a single replicate does not provide an appropriate estimate of the experimental error, nor does repeated measurement of the experimental units. Estimates may be obtained by measurement of multiple units within each treatment or by replication of the basic experiment in time or space. Under certain assumptions a single replicate (or even a partial replicate) of a large factorial experiment may provide a satisfactory estimate of the experimental error. However, it is suggested that a statistician be consulted if such an approach is being considered.

Perhaps the most intuitive concept of replication is that increasing replication increases the ability to detect real differences (sensitivity). What is perhaps not so obvious to most researchers is the rapidly decreasing benefit from additional replication (concept of diminishing returns). Roughly, it requires a 4-fold increase in replication to double the sensitivity of an experiment (detect a difference of one-half the size). If we arbitrarily say that 1 replicate has a sensitivity of 1.0, then 2, 4, 16, and 100 replicates would have sensitivities of 1.4, 2.0, 4.0, and 10.0, respectively. Good experimental design is also concerned with over-replication since it results in a waste of resources. Remember, however, that the minimum replication is 2 for estimation of the experimental error.

It is frequently desirable to test treatments over a diverse set of experimental units or conditions if general conclusions are to be drawn from the experiment. For example, the agronomist would like the results of his experiment to be true for a variety of soil types, levels of fertility, and environmental conditions (years). However, if he finds that the results are not the same under all conditions, it is important to know which conditions produce different results. When experiments of this nature are conducted, it is important that the concept of local control, which will be discussed in a later section, be used to prevent serious inflation of the experimental error due to the lack of uniformity of the experimental units.

Randomization

Replication makes the estimation of the experimental error possible, while it is randomization that assures the validity of the estimate. In addition, randomization assures that unbiased estimates of the differences between treatment means are obtained. It is not that we can assure that every treatment (or experimental unit within a treatment) is treated identically. Rather, it is a concept of fairness. That is, every treatment should have an equal probability of being assigned to the more or less desirable experimental units.

Randomization requires that some technique of chance be used, whether it be flipping a coin, tossing a die, drawing numbered tags, or using a random number table. Random does not mean systematic, uniform, or haphazard. Although a systematic arrangement may at times result in more accurate estimates of the treatment differences, its use cannot be justified. When systematic procedures are used, no reliable estimate of the experimental error is available and thus meaningful inferences are not possible.

Simple random assignment of treatments to the experimental units is not sufficient. By oversight or lack of understanding of the principle of randomization, many researchers fail to use randomization at other stages of their experiment. For example, if at any stage during the experiment when the assignment of units to some facility (or instrument), the order of processing units, or the order of measurement or collection of data may influence the results, then the order should be randomized in a manner consistent with the design of the experiment.

Local Control

Local control is to a large extent what the selection of an experimental design is all about. Local control refers to techniques (design, analysis, selection, or refinement of laboratory analysis) that are used to reduce (or control) the experimental error. The most widely used design technique is known as blocking.

Blocking is the grouping, prior to the assignment of treatments, of experimental units into homogeneous groups, known as blocks. The concept is to assign blocks in such a manner as to maximize the difference between blocks while minimizing the differences within blocks. Treatments are then assigned at random within blocks. Thus, the observed differences between treatments within blocks are largely due to treatment effects, and are less likely to be due to random variation. The differences between blocks are not included in the experimental error, thus reducing the magnitude of the error variance.

Experimental units may simultaneously be blocked for more than one source of heterogeneity. While the simplest one-directional blocking design is called the randomized complete block, a two-directional blocking design is the Latin square. The application and analysis of these designs are presented in Chapter 8. Many other designs have been developed to take advantage of the blocking concept: incomplete blocks, balanced incomplete blocks, Greco-Latin squares, lattices, to name a few. The use of blocking and how to block generally requires a good amount of intuition and practical experience. The statistician can generally do little more than provide the researcher with the principles of blocking, unless he has had some prior experience himself.

The most common analysis technique used to reduce the experimental error is the Analysis of Covariance. This analysis makes use of an "indicator variable(s)" that provides information about variability that may have existed in the experimental units prior to (or at least independent of) treatment. Since this variability existed prior to the study or is at least independent of the treatment, this portion of the variability is removed from the experimental error.

Sensitivity

Sensitivity is defined here as the ability to detect real differences if they exist. That is, given two experimental designs, the one that can detect the smaller difference between two means has the greater sensitivity. In many texts, the terms sensitivity, precision, efficiency, and amount of information have very similar meanings. We have discussed the concept of sensitivity in connection with replication, namely, that a 4-fold increase in replication approximately doubles the sensitivity of an experiment. In this connotation, sensitivity is being defined as $1/S_{\bar{Y}}$, where $S_{\bar{Y}} = (\sigma^2/n)^{1/2}$. Thus, sensitivity proportionally increases as the square root of n increases, and as the standard deviation decreases (square root of the experimental error).

It should be noted that, in small experiments, the increase in error degrees of freedom from increasing n also increases the ability to detect smaller differences. Although we generally think of sample size and experimental error as being independent, this is not always the case. As sample size increases, the researcher may be forced to use less uniform experimental material to meet his needs. Some increase in experimental error is likely to occur when diverse experimental units are used to provide greater scope of inference, even with the use of local control. One can also expect an increase in experimental error as the number of treatment combinations increases, since blocking techniques will become less effective with greater block size.

Confounding

The word confounding simply means the mixing together of effects. When it is impossible (generally because of design) to separate the effects in the statistical analysis, we say they are confounded. Unintentionally confounding may result in portions (or the total) of an experiment yielding ambiguous results. For example, let us consider again the cattle feeding experiments presented earlier in this chapter.

We have already discussed the lack of replication in this experiment, now let's turn our attention to a second problem. It would be rare, in spite of his or her best efforts, that the researcher could create exactly uniform conditions. Therefore, it is likely that the 3 pens used in the experiment are not equally favorable. If the experiment were to be analyzed, treatment and pen effects would be inseparable, or "confounded." If the researcher concluded that there were differences between treatment groups, it would be just as easy to argue that the differences were due to pen effects as it would be to argue that they were due to treatment effects.

It may, however, be surprising to many that confounding is intentionally used in some experimental designs. The researcher may intentionally confound effects which are irrelevant for the interpretation of his experiment. For example, blocking effects are frequently the result of confounding many different effects, rather than a single source of variation. In factorial experiments, with many factors, it is frequently safe to assume that higher order interactions are negligible and may, therefore, be confounded with blocks or certain other treatment factors. In most cases, the reasons for confounding as a design technique are to reduce block size, thus permitting greater uniformity within block and a reduction in the experimental error, and an overall reduction in the size of the experiment.

Orthogonality

When differences between arithmetic treatment means are due only to the single desired treatment effect plus random error, then the treatment effects are said to be orthogonal. The concept of orthogonality is important in both design and analysis. Nonorthogonality may originate from the design used or from unequal replication of treatments (or unbalanced).

The following example illustrates the concept of nonorthogonality. Suppose an experiment is conducted with experimental units of 2 sexes subjected to 2 treatments with recorded responses on 6 experimental units. The data are given in the following table with computed sex and treatment means.

| | Treatment | | |
Sex	A	B	Mean
M	10,8	6	8
F	6	4,2	4
Mean	8	4	

The researcher concludes that the response was 4 units greater in males than females, and 4 units greater for treatment A than for B. The researcher then decides to examine the 4 treatment means and constructs the following table of means.

Sex	Treatment	
	A	B
M	9	6
F	6	3

He now notes that in both treatments males are only 3 (not 4) units greater than females, and that for both sexes treatment A is only 3 (not 4) units greater than treatment B. The apparent discrepancy is due to the nonorthogonality of the treatment and sex effects, resulting in the confounding of those effects. Since more males were assigned to treatment A, while more females were assigned to treatment B, part of the observed differences between sex means and between treatment means in the first table were in fact due to the other effect; that is, males only *appeared* to be 4 units greater (instead of 3) because part of the difference was due to the greater response of treatment A, where most of the male experimental units were assigned. A similar argument could also be presented with respect to the 4 unit treatment response.

The presence of nonorthogonality due to design or unequal replication requires careful interpretation. The researcher should seek statistical assistance before intentionally using such a design, as well as advice concerning appropriate procedures for analysis.

REFERENCES

For additional information on experimental design:

COCHRAN, W.G. and COX, G.M.   1957.   Experimental Design, 2nd Edition. John Wiley & Sons, New York.

COX, D.R.   1958.   Planning of Experiments.   John Wiley & Sons, New York.

FEDERER, W.T.   1955.   Experimental Design.   Macmillan Company, New York.

KEMPTHORNE, O.   1952.   The Design and Analysis of Experiments. John Wiley & Sons, New York.

LI, C.C.   1964.   Introduction to Experimental Statistics.   McGraw-Hill Book Company, New York.   Chapters 9–28.

WINER, B.J.   1962.   Statistical Principles in Experimental Design, 2nd Edition.   McGraw-Hill Book Company, New York.

For information on sampling designs:

BABBIE, E.R.   1973.   Survey Research Methods.   Wadsworth Publishing Company, Belmont, Calif.

7

Analysis of Variance I: The 1-Way ANOVA

In Chapter 5 we were concerned with comparisons involving two samples. The techniques presented were used to compare the mean of a treated sample vs the mean of a control sample or to compare the means of two treated samples. The more common case in biological research is to include several treatment groups in the same experiment. Although the novice may suggest using multiple t-tests, such an approach is neither appropriate nor computationally simple when the number of groups being tested is large.

In this chapter we introduce the technique known as the analysis of variance (abbreviated ANOVA, ANOV or AOV). We may view the ANOVA as a technique to test whether two or more sample means belong to populations with identical parametric means. Where only two samples are compared, the results of the ANOVA are identical to the results of the t-test. You can view the ANOVA as a more general procedure than the t-test and, as you become more familiar with the technique, you are likely to make use of the ANOVA even when the t-test is appropriate.

THE LINEAR ADDITIVE MODEL

Suppose we wish to examine 4 samples to test the hypothesis that they were drawn from populations with identical means. The problem then is to examine the response variable (Y) and determine if there is sufficient evidence to reject the null hypothesis of no difference between the population means (H_o: $\mu_1 = \mu_2 = \mu_3 = \mu_4$). Now suppose that we look at only the first replicate of the experiment and find the following results:

Sample	1	2	3	4
Response	59	37	55	60

Examination of the data might lead some readers to conclude that differences in population means do exist, while other readers would appropriately conclude that more information is needed before any conclusion can be reached. What additional information is needed to draw conclusions about differences between population means? Although the variance among samples appears to be large, we would like to know how variable the response is within samples.

Table 7.1 presents all the data for the experiment. It is clear from examination of all the data that there is considerable variability within the samples. What is of concern is how great the variability is among samples compared to the variability within samples. The greater the variability among samples relative to the variability within samples the stronger the evidence supporting a difference between sample means. The ANOVA is a technique which compares the among and within sample variation in order that conclusions may be drawn about differences among the population means.

TABLE 7.1. SAMPLE DATA SET FOR THE COMPLETELY RANDOM DESIGN

Replicates (n)	Sample Group (a)				
	1	2	3	4	
1	59	37	55	60	
2	49	49	68	43	
3	63	58	54	53	
Sums ($Y_i.$)	171	144	177	156	$Y.. = 648$
Mean ($\bar{Y}_i.$)	57	48	59	52	$\bar{Y}.. = 54$
Variance (S_i^2)	52	111	61	73	

Before we can present the ANOVA in a formal way, we need to develop appropriate notation for handling the data and presenting the computations. In Table 7.2 we have presented in symbolic notation the data to be analyzed by 1-way ANOVA. The data are classified in two directions; therefore, an individual observation would be identified as a Y_{ij}, where Y represents the numeric value of the measured response, while i and j identify the sample group and replication number, respectively. For example, Y_{23} would represent the observation for the second sample, third replicate. In Table 7.1 we see that $Y_{23} = 58$.

When multiple subscripts are required, summation notation becomes considerably more cumbersome. In summation notation, the sum of all the observations would be denoted $\sum_{i=1}^{a} \sum_{j=1}^{n} Y_{ij}$, and the sample sums would be denoted $\sum_{j=1}^{n} Y_{ij}$, or the sum for sample 2 would be $\sum_{j=1}^{n} Y_{2j}$. Although this may not seem unwieldy for the 1-way ANOVA, you can probably appreciate the difficulty if we were dealing with data classified in 4 ways

$$\left(\sum_{i=1}^{a} \sum_{j=1}^{b} \sum_{k=1}^{c} \sum_{l=1}^{n} Y_{ijkl} \right)$$

TABLE 7.2.   SYMBOLIC DATA ARRANGEMENT AND SUMS FOR THE 1-WAY ANALYSIS OF VARIANCE

		Sample Group			
	Replicate	1	2	...	a
	1	Y_{11}	Y_{21}	...	Y_{a_1}
	2	Y_{12}	Y_{22}	...	Y_{a_2}
	.	.	.		.
	.	.	.		.
	.	.	.		.
	n	Y_{1n}	Y_{2n}	...	Y_{an}
Sums ($\sum\limits_{j=1}^{n} Y_{ij} = Y_{i.}$)		$Y_{1.}$	$Y_{2.}$	...	$Y_{a.}$	$Y_{..} = \sum\limits_{i=1}^{a} Y_{i.}$
Means ($\bar{Y}_i = \bar{Y}_{i.}$)		$\bar{Y}_{1.}$	$\bar{Y}_{2.}$	...	$\bar{Y}_{a.}$	$\bar{Y}_{..} = \Sigma Y_{ij}/n$

In order to simplify the task of writing the multiple summation notation we have adopted what has sometimes been referred to as "dot" notation. In "dot" notation, to indicate that summation has taken place, we simply replace the appropriate subscripts with a dot. For example, $\sum\limits_{i=1}^{a} \sum\limits_{j=1}^{n} Y_{ij}$ would be denoted $Y_{..}$ and sample sums $\sum\limits_{j=1}^{n} Y_{ij}$ would be $Y_{i.}$ (the sum of sample 2 would be $Y_{2.}$).

In addition, the dot notation is useful in computing sums of squares. The dot is used to indicate summation which has taken place before squaring while the capital sigma (Σ) indicates summation after squaring. The following equalities will help illustrate the use of the dot notation:

$$\sum\limits_{i=1}^{a} \sum\limits_{j=1}^{n} Y_{ij} = Y_{..}$$

$$\sum\limits_{j=1}^{n} Y_{ij} = Y_{i.}$$

$\sum\limits_{i=1}^{a} \left(\sum\limits_{j=1}^{n} Y_{ij} \right)^2 = \sum\limits_{i=1}^{a} Y^2_{i.}$ or $\Sigma Y^2_{i.}$ since the summation is assumed to be from the first to the last if the limits above and below the sigma are omitted.

$$\sum\limits_{i=1}^{a} \sum\limits_{j=1}^{n} Y^2_{ij} = \Sigma Y^2_{ij}$$

$$\left[\sum\limits_{i=1}^{a} \sum\limits_{j=1}^{n} Y_{ij} \right]^2 = Y^2_{..}$$

Before the analysis of variance can be used to make statistical inferences about differences among population means, certain assumptions must be made about the observed data. The statistical model for the analysis of variance assumes a linear model in which the effects are additive and the observations are independent and drawn from normal distributions with equal variances.

In a formal way we would state the linear additive model for the 1-way ANOVA as:

$$Y_{ij} = \mu + \alpha_i + \epsilon_{ij}$$

where $i = 1, \ldots, a$
$j = 1, \ldots, n$
μ is the true mean
α_i is the true deviation of the mean of the ith group from μ
and ϵ_{ij} is the random deviation of the jth observation of the ith group from the mean of the ith group

These deviations (ϵ_{ij}) are assumed to be normally and independently distributed with a mean of zero and a common variance [NID $(0, \sigma^2)$].

Simply, the linear additive model says that the value of each observation (Y_{ij}) is the sum of 3 effects. The first is the grand mean (μ); the second is how much the group mean (μ_i), to which Y_{ij} belongs, deviates from the grand mean; and the third is how much the observation (Y_{ij}) deviates from its group mean (μ_i). The data of Table 7.1 provide estimates of these effects. The 3 effects are estimated as (the "hat" symbol $\hat{\ }$ means "the estimate of," i.e., $\hat{\mu}$ means "estimate of μ"):

$$\hat{\mu} = \bar{Y}_{..}$$
$$\hat{\alpha}_i = (\bar{Y}_{i.} - \bar{Y}_{..})$$
$$\hat{\epsilon}_{ij} = (Y_{ij} - \bar{Y}_{i.})$$

For example, if we were to estimate these effects for observation Y_{11} of Table 7.1, we would find:

$$\hat{\mu} = 54$$
$$\hat{\alpha}_1 = (57 - 54) = +3$$
and $$\hat{\epsilon}_{11} = (59 - 57) = +2$$

That is, $Y_{11} = 54 + 3 + 2 = 59$. Estimates of the effects for all of the observations are given in Table 7.3. Note that since the effects are defined

TABLE 7.3. DECOMPOSITION OF OBSERVED VALUES INTO ESTIMATES OF EFFECTS FOR 1-WAY ANOVA

	Sample Group			
	1	2	3	4
	$\mu + \alpha_1 + \epsilon_{1j}$	$\mu + \alpha_2 + \epsilon_{2j}$	$\mu + \alpha_3 + \epsilon_{3j}$	$\mu + \alpha_4 + \epsilon_{4j}$
	54 + 3 + 2	54 − 6 − 11	54 + 5 − 4	54 − 2 + 8
	54 + 3 − 8	54 − 6 + 1	54 + 5 + 9	54 − 2 − 9
	54 + 3 + 6	54 − 6 + 10	54 + 5 − 5	54 − 2 + 1
Mean	54 + 3 + 0	54 − 6 + 0	54 + 5 + 0	54 − 2 + 0
Grand Mean		54 + 0 + 0		

as deviations, the estimates of ϵ_{ij} have a mean of zero for each group; and α_i and ϵ_{ij} have means of zero for the total data set.

Although we have specifically stated that the ϵ_{ij} are random effects, nothing has been said about the α_i. The description of the model would not be complete without further classification of the α_i. The researcher has to identify the α_i as being either random or fixed effects. If the researcher chooses the random effects model, he assumes that the α_i are NID $(0, \sigma_\alpha^2)$. The selection of this model implies that the researcher is concerned with some population of possible groups from which sample groups are randomly selected to be included in the experiment. Generally, the researcher's main interest in the random effects model is the estimation of the variance components σ^2 and σ_α^2, rather than the estimation and testing of differences between group means. When the researcher chooses the fixed effects model $(\Sigma\alpha_i = 0)$, he is in effect saying that he is concerned only with the groups included in the experiment. These a sample groups were selected by the researcher (not randomly) to be included in the experiment because he wished to estimate differences between them and/or to test specific hypotheses about the group means. This model is referred to as the fixed model or Model I, while the model above is known as the random model or Model II.

THE COMPLETELY RANDOM DESIGN

When the treatments (sample groups) are assigned to the experimental units by a completely random procedure (except for the possible restriction that an equal number of experimental units be assigned to each treatment), the resulting design is referred to as the completely random design (CRD). The use of the CRD implies that the experimental units were relatively uniform or at least no other assignable source of variation could be identified. The only source of variation that is assignable is that due to treatment (sample groups). All other variation is considered to be experimental error and as such, is included in our estimate of population variance (σ^2). The use of the CRD also implies that any action taken during the study which might result in differences among experimental units also requires randomization of the order in which the action is carried out: for example, the assignment of experimental animals to pens, the order in which plots are harvested, or how samples are sequenced to carry out laboratory analysis.

Suppose that Table 7.1 represents data from a CRD. As stated earlier we need to compare the variance among sample groups with the variance within sample groups. Now let us look at how these two variances can be computed. First, consider the estimate of the within group variance.

Each S_i^2 in Table 7.1 estimates the within population variance, if our assumption of a common variance is true. What we wish to obtain is a single estimate of the within population variance. Since the amount of information in a sample variance is indicated by the number of degrees of freedom, an average variance weighted for the degrees of freedom is computed.

$$S_W^2 = \frac{(n_1-1)S_1^2 + (n_2-1)S_2^2 + \ldots + (n_a-1)S_a^2}{(n_1-1) - (n_2-1) + \ldots + (n_a-1)}$$

$$= \frac{(3-1)(52) + (3-1)(111) + (3-1)(61) + (3-1)(73)}{(3-1) + (3-1) + (3-1) + (3-1)}$$

$$= 74.25$$

This variance then is referred to as the within group variance and is an estimate of the true population variance σ^2 One should note that since the within variance is computed from deviations between each observation and its own group mean, differences between groups should not in any way affect S_W^2. within group SS.

Now we turn our attention to the estimation of the variance among sample groups. As a measure of the variability among groups, suppose we treat the 4 group means as if they were 4 observations and compute the variance.

$$S_{\bar{Y}}^2 = \frac{\Sigma(\bar{Y}_{i\cdot} - \bar{Y}_{\cdot\cdot})^2}{a-1}$$

$$= \frac{(57-54)^2 + (48-54)^2 + (59-54)^2 + (52-54)^2}{4-1}$$

$$= 24.667$$

If all samples were in fact drawn from populations with the same mean (and common variances) then $S_{\bar{Y}}^2$ would be an estimate of $\sigma_{\bar{Y}}^2$ computed from the 4 sample means. This estimate of the variance cannot be directly compared with S_W^2, since $S_{\bar{Y}}^2$ is based on the variance among the means (i.e., $\bar{Y}$'s), rather than individual observations. In Chapter 4, we learned that $S_{\bar{Y}}^2$ could be computed from a single sample of n observations by the equation:

$$S_{\bar{Y}}^2 = \frac{S^2}{n}$$

$$\therefore S^2 = nS_{\bar{Y}}^2$$

From these equations we can see that if population variance is to be estimated from means, we must multiply the variance of the means by the size of the sample. When population variance is computed in this manner, it is referred to as the variance among groups and denoted S_A^2. For the data of Table 7.1, we find

$$S_A^2 = n \, S_{\bar{Y}}^2$$

$$= 3 \, (24.667)$$

$$= 74.00$$

We now have two independent estimates of the population variance which can be compared by the F statistic introduced in Chapter 5. If the 4 samples were drawn from populations with identical means, then $S_A^2 \cong S_W^2$, and the F ratio $S_A^2/S_W^2 \cong 1$. However, should the samples be drawn from populations with different means (i.e., treatment effects), then S_A^2 should be greater, while S_W^2 should not be affected; therefore, the F ratio S_A^2/S_W^2 should be greater than 1. For the data of Table 7.1 the F ratio is $F = \dfrac{S_A^2}{S_W^2} = \dfrac{74.00}{74.25} = 0.997$. Since S_A^2 was based on 4 group means, the numerator has 3 degrees of freedom. Since S_W^2 was based on 4 variances with 2 degrees of freedom each, the denominator has 8 degrees of freedom. The 5% tabular F value (Appendix Table D.1) is 4.07. That is, based on ratios of variance with 3 and 8 degrees of freedom, the observed F ratio would have to be greater than or equal to 4.07 before we would reject the null hypothesis of equal population means. Therefore, we would accept H_o, since $0.997 < 4.07$.

To simulate treatment effects, suppose that we add or subtract a constant from each of the observations in each group of Table 7.1, thus creating the new set of observations in Table 7.4. Note that the sample variances have not changed; therefore, $S_W^2 = 74.25$ also remains unchanged.

TABLE 7.4.   SAMPLE DATA SET WITH SIMULATED GROUP EFFECTS

Replicates (n)	Sample Group (a)				
	1(+0)	2(−5)	3(+9)	4(−4)	
1	59	32	64	56	
2	49	44	77	39	
3	63	53	63	49	
Mean ($\bar{Y}_{i\cdot}$)	57	43	68	48	$\bar{Y}_{\cdot\cdot} = 54$
Variance (S_i^2)	52	111	61	73	

However, we must compute S_A^2 for the new data set.

$$S_{\bar{Y}}^2 = \frac{\Sigma(\bar{Y}_{i\cdot} - \bar{Y}_{\cdot\cdot})^2}{a - 1} = \frac{362}{4 - 1} = 120.667$$

$$S_A^2 = nS_{\bar{Y}}^2 = 3 \, (120.667) = 362$$

$$F = \frac{S_A^2}{S_W^2} = \frac{362}{74.25} = 4.88$$

Since the calculated $F = 4.88$ is greater than the tabular $F (4.07)$, there is sufficient evidence to conclude that the population means are not equal. The variance within (S_W^2) is an estimate of the population variance (σ^2). This relationship is generally not a difficult one to understand; however, it is not as obvious what the "among" variance (S_A^2) estimates. When group effects are absent, S_A^2 is shown to be an independent estimate of the population variance (σ^2). If group effects are present, then S_A^2 would be expected to be an estimate of population variance plus some added component due to group effects. It can be shown that S_A^2 is an estimate of

$$\sigma^2 + n\left(\frac{\Sigma\alpha_i^2}{a-1}\right)$$

The term $\dfrac{\Sigma\alpha_i^2}{a-1}$ is a quantity which looks very much like a variance, since the α_i are defined as deviations. In the case of the fixed effects model (Model I), this quantity is the variance of the selected treatment groups, but should not be considered an estimate of some population variance. For the random effects model (Model II), $\dfrac{\Sigma\alpha_i^2}{a-1}$ is an estimate of σ_α^2. The quantity σ_α^2 is referred to as a variance component and is a measure of the added variance due to the difference among the randomly selected groups.

The expected F ratios for the Model I and Model II ANOVA are $\dfrac{\sigma^2 + n\left(\dfrac{\Sigma\alpha^2}{a-1}\right)}{\sigma^2}$ and $\dfrac{\sigma^2 + n\sigma_\alpha^2}{\sigma^2}$, respectively. Examination of these expected F ratios make it obvious why the ANOVA is in fact a test of differences among groups and why in the absence of group effects the expected F ratio equals 1. When true group differences are zero, the $\Sigma\alpha^2$ (or σ_α^2) would equal zero and the expected F ratio would be $\dfrac{\sigma^2}{\sigma^2} = 1$. On the other hand, when true group differences exist (i.e., when the μ_i are not equal), then $\Sigma\alpha^2$ (or σ_α^2) would be greater than zero and the expected F ratio would be greater than 1.

In application, the problem is to decide how much greater than 1 the F statistic must be to support rejection of the null hypothesis of equality of the population group means. This decision is made by comparison of the sample F ratio with the tabular F ratio at the level of significance determined by the researcher and with degrees of freedom associated with the S_A^2 and S_W^2. That is, any sample F ratio that is less than the tabular F ratio is consistent with the null hypothesis of no group differences at the stated level of significance.

ANOVA—EQUAL REPLICATION

The analysis of variance is generally conducted in a systematic manner using computing formulas (rather than the more intuitive approach followed in the previous sections) and presented in tabular form. In this section, we will apply these formulas to a couple of data sets and present the results in a standard analysis of variance table.

Table 7.5 presents both the definitional and computing formulas for the sums of squares of the ANOVA. Note that computational formulas are based on the Y_{ij} and various totals. It is not necessary to compute means or individual group variances as was done in the previous section to complete the analysis of variance. Also note that only 2 of the 3 sums of squares (SS) need to be computed from the data to complete the ANOVA. The remaining one, computed by subtraction, is generally the within SS since it is the most difficult to compute directly. The definitional formulas are given to assist the student with his or her thinking about what variability each source is measuring. It is difficult for the novice to visualize what is being computed if only the ANOVA computing formulas are presented.

TABLE 7.5.   DEFINITIONAL AND COMPUTING SUMS OF SQUARES FOR THE 1-WAY ANOVA

Source of Variation	Degrees of Freedom	Sum of Squares Definitional	Sum of Squares Computing
Among groups	$a - 1$	$n \Sigma(\bar{Y}_{i\cdot} - \bar{Y}_{\cdot\cdot})^2$	$(\Sigma Y_i^2/n) - (Y^2_{\cdot\cdot}/an)$
Within groups	$a(n - 1)$	$\Sigma (Y_{ij} - \bar{Y}_{i\cdot})^2$	by subtraction
Total	$an - 1$	$\Sigma (Y_{ij} - \bar{Y}_{\cdot\cdot})^2$	$\Sigma Y^2_{ij} - (Y^2_{\cdot\cdot}/an)$

Table 7.6 is the complete symbolic ANOVA table. The phrase "mean squares" is simply another statistical term for variance. The term mean squares comes from the definition of population variance of "average squared deviation from the mean." Therefore, MS_A is the variance between groups (S_A^2) and MS_W is the variance within groups (S_W^2). We are now ready to proceed with a numeric example of the analysis of variance table. So that we may compare the results here with the computations in the last section let us first analyze the data of Table 7.1. The necessary totals are given in the table and the sums of squares would be computed as follows:

TABLE 7.6.   SYMBOLIC 1-WAY ANOVA FOR THE COMPLETELY RANDOM DESIGN

Source of Variation	Degrees of Freedom	Sum of Squares	Mean Squares (or Variance)	F Ratio
Among groups	df_A	SS_A	$MS_A = SS_A/df_A$	MS_A/MS_W
Within groups	df_W	SS_W	$MS_W = SS_W/df_W$	—
Total	df_T	SS_T	—	—

$$SS_T = \Sigma Y_{ij}^2 - \frac{Y_{..}^2}{an}$$

$$= (59^2 + 49^2 + \ldots + 53^2) - \frac{648^2}{(4)(3)}$$

$$= 816$$

$$SS_A = \frac{\Sigma Y_{i.}^2}{n} - \frac{Y_{..}^2}{an}$$

$$= \frac{171^2 + 144^2 + \ldots + 156^2}{3} - \frac{648^2}{(4)(3)}$$

$$= 222$$

$$SS_W = SS_T - SS_A$$

$$= 816 - 222$$

$$= 594$$

The complete ANOVA table would be as follows:

Source of Variation	df	SS	MS	F
Among groups	3	222	74.00	0.997
Within groups	8	594	74.25	
Total	11	816		

The values of MS_A, MS_W, and F are identical to those presented in the previous section. However, this is not always the case. The computational formulas are not only easier to work with, but are also subject to less rounding error, thus frequently resulting in more precise values.

Now, consider a set of data from an experiment designed to test for differences among 4 dietary treatments. Later in the chapter we will consider in more detail what dietary treatments were used. But for the time being let us simply say that the 4 treatments were selected by the researcher, thus suggesting the fixed model ANOVA. The null hypothesis to be tested is that the true treatment means are equal (H_o: $\mu_1 = \mu_2 = \mu_3 = \mu_4$), while the alternative hypothesis is that at least 1 μ is different.

A completely random design with 6 beef steers per treatment was used to test the hypothesis. The variable of interest was the weight gained (kg) during a 60 day feeding trial. The resulting data and the necessary summations are presented in Table 7.7. Table 7.8 presents the computations for the sums of squares and the complete ANOVA table.

TABLE 7.7. WEIGHT GAINS (KG) OF 24 BEEF STEERS ON 4 DIETS DURING A 60 DAY FEEDING TRIAL

		Diet		
	1	2	3	4
	91	110	85	80
	89	84	92	72
	102	84	89	89
	82	85	85	79
	93	89	72	72
	93	87	80	76
$Y_i.$	550	539	503	468

$Y.. = 2060$

TABLE 7.8. COMPUTATION OF SUMS OF SQUARES AND THE ANOVA FOR THE DATA OF TABLE 7.7

$$SS_T = 91^2 + 89^2 + \ldots + 76^2 - \frac{2060^2}{24}$$

$$= 1863$$

$$SS_A = \frac{550^2 + 539^2 + 503^2 + 468^2}{6} - \frac{2060^2}{24}$$

$$= 692$$

$$SS_W = 1863 - 692 = 1171$$

Source of Variation	df	SS	MS	F
Among diets	3	692	230.7	3.94
Within diets	20	1171	58.55	—
Total	23	1863	—	—

The sample F ratio of 3.94 would be compared to the tabular F value with 3 degrees of freedom in the numerator and 20 degrees of freedom in the denominator. The 0.05 tabular $F_{0.05(3,20)}$ equals 3.10. Therefore, we would reject the null hypothesis ($P < 0.05$) and conclude that the treatment means are not equal.

The reader should note that we reached this conclusion and had not yet computed the treatment means. The ANOVA does not identify which mean(s) is (are) different. The determination of specific differences will be considered in the last section of this chapter. The purpose of the ANOVA is simply to provide evidence concerning the presence or absence of treatment effects.

ANOVA—UNEQUAL REPLICATION

With a few changes in the computing formulas for sums of squares and the formulas for degrees of freedom, the 1-way ANOVA may be completed for data sets with unequal replication. The appropriate formulas are given in Table 7.9. The change in notation concerns the number of replications. Since n is not the same for every group, the number of replications is denoted n_i, where i corresponds to the group identification.

TABLE 7.9. FORMULAS FOR DEGREES OF FREEDOM AND COMPUTATION OF SUMS OF SQUARES FOR UNEQUALLY REPLICATED 1-WAY ANOVA

Source of Variation	df	Computing SS
Among groups	$a - 1$	$\Sigma(Y_i^2/n_i) - (Y^2/n.)$
Within groups	$n. - a$	by subtraction
Total	$n. - 1$	$\Sigma Y_{ij}^2 - (Y^2/n.)$

To illustrate the 1-way ANOVA procedure for unequal replication, suppose that we wish to examine the yearling height of colts from 7 thoroughbred sires. The data and the ANOVA are presented in Tables 7.10 and 7.11, respectively. If we assume that the sires used in the experiment represent a random sample of thoroughbred sires, then we are dealing with a random effects model. Note that although the example used for equal replication was a fixed model and that the example here is a random model, the type of model used has nothing to do with equal or unequal replication.

TABLE 7.10. YEARLING HEIGHTS IN HANDS FOR COLTS SIRED BY 7 THOROUGHBRED STALLIONS

Stallion	1	2	3	4	5	6	7
	12.5	13.5	12.0	14.5	12.0	13.0	13.0
	13.0	11.5	10.5	13.5	12.0	12.0	10.0
	11.0		11.0	13.0	11.5	10.5	9.5
	12.5		11.5			12.5	12.0
			11.0				
			10.0				
$Y_i.$	49.0	25.0	66.0	41.0	35.5	48.0	44.5
n_i	4	2	6	3	3	4	4

$$Y.. = 309.0$$
$$n. = 26$$

Since the sample F ratio is greater than the tabular F value ($F_{0.05(6,19)} = 2.63$), we would conclude that sire effects are present. Generally, in the case of a random model, the interest of the researcher is not so much in the test of significance, but rather in the estimation of the added variance component due to the random group effect. Earlier in this chapter we found that for the random model $MS_W \doteq \sigma^2$ and $MS_A \doteq \sigma^2 + n\sigma_\alpha^2$, where σ_α^2 in this case is the added variance component due to sires ($\doteq$ is read "is an estimate of . . . .").

TABLE 7.11.   COMPUTATION OF SUMS OF SQUARES AND ANOVA FOR DATA OF TABLE 7.10

$$SS_Y = \Sigma Y^2_{ij} - \frac{Y_{..}^2}{n\cdot}$$

$$= 12.5^2 + 13.0^2 + \ldots + 12.0^2 - \frac{309.0^2}{26}$$

$$= 37.65$$

$$SS_A = \Sigma \left[\frac{Y^2_{i\cdot}}{n_i} \right] - \frac{Y_{..}^2}{n}$$

$$= \frac{49.0^2}{4} + \frac{25.0^2}{2} + \ldots + \frac{44.5^2}{4} - \frac{309.0^2}{26}$$

$$= 17.88$$

$$SS_W = SS_T - SS_A$$

$$= 37.65 - 17.88$$

$$= 19.77$$

Source of Variation	df	SS	MS	F
Among sires	6	17.88	2.98	2.87
Within sires	19	19.77	1.04	—
Total	25	37.65	—	—

We could now, by simple algebra, solve for the sire component of variance if we had an equally replicated (n) experiment. For the unequally replicated experiment we must substitute n_0 for n in the equation where n_0 is a kind of average computed as follows:

$$n_0 = \frac{1}{a-1}\left[n\cdot - \frac{\Sigma n_i^2}{n\cdot} \right]$$

For our example n_0 equals:

$$= \frac{1}{7-1}\left[26 - \frac{4^2 + 2^2 + \cdots + 4^2}{26} \right]$$

$$n_0 = 3.65$$

The estimate of the sire component of variance would be:

$$S_\alpha^2 = \frac{MS_A - MS_W}{n_0}$$

$$= \frac{2.98 - 1.04}{3.65}$$

$$= 0.53$$

Although the fixed effects model is by far the most common application of the 1-way ANOVA, the estimation of variance components like the last example is used by geneticists in determining the importance of heredity for a given trait. Variance components are also of value to researchers who are attempting to design experiments to maximize efficiency of their research effort. For the random model, unlike the fixed model, the researcher will probably not be concerned with identifying group differences since the groups (sires) were selected at random.

COMPARISONS BETWEEN MEANS

In this section we will consider the problem of how to complete the analysis for the fixed model. Although there are numerous comparison procedures, only three procedures will be considered here. The three presented are commonly used in agriculture and the food sciences.

Orthogonal Linear Comparisons

The most sensitive technique is "orthogonal linear comparisons." A comparison is simply a linear function of the treatment means (or totals) which examines the differences among the treatments. Orthogonal simply means that the comparisons, if more than one is computed, measure independent effects.

The equation for a linear comparison is $\Sigma c_i \bar{Y}_i$, where the $\Sigma c_i = 0$. The c_i are constants which define how the treatments are to be compared.

To illustrate the comparison concept, consider the feeding trial experiment presented earlier in the chapter. Suppose that the 4 dietary treatments were different protein (or nitrogen) sources.

Treatment	Mean
1—Soybean meal	91.67
2—Cottonseed meal	89.83
3—Urea	83.83
4—Ammonia	78.00

Given these treatments, 3 comparisons are suggested.
Comparison 1—Soybean *vs* Cottonseed

$$c_1 = (1)(\bar{Y}_1.) + (-1)(\bar{Y}_2.)$$

where the 1 and -1 are the constants used to define the contrast. For this comparison the c_i are defined such that the value of the comparison is the difference between the means.

$$c_1 = (1)(91.67) + (-1)(89.83) = 1.84$$

The sum of squares for a comparison is defined as follows;

$$SS_1 = \frac{n(\Sigma c_i \bar{Y}_i)^2}{\Sigma c_i^2}$$

$$= \frac{6\,(1.84)^2}{1^2 + (-1)^2}$$

$$= 10.2$$

Comparison 2—Urea *vs* Ammonia

$$c_2 = (1)(83.83) + (-1)(78.00)$$

$$SS_2 = \frac{6(5.83)^2}{2} = 102.0$$

Comparison 3—Protein *vs* Nonprotein Nitrogen Source

$$c_3 = (\tfrac{1}{2})(91.67) + (\tfrac{1}{2})(89.83) + (-\tfrac{1}{2})(83.83) + (-\tfrac{1}{2})(78.00)$$

$$= 9.84$$

$$SS_3 = \frac{6(9.84)^2}{(\tfrac{1}{2})^2 + (\tfrac{1}{2})^2 + (-\tfrac{1}{2})^2 + (-\tfrac{1}{2})^2}$$

$$= 581.0$$

Now consider the question of orthogonality. No more than a − 1 comparisons may be orthogonal. The sum of the sums of squares for a − 1 orthogonal comparisons will equal the among treatment sums of squares (SS_A) from the ANOVA. Any 2 comparisons are orthogonal when $\Sigma c_{1} c_{2}$ equals zero, where 1 and 2 are any 2 comparisons. For our comparisons we find:

$$\Sigma c_{1} c_{2} = (1)(0) + (-1)(0) + (0)(1) + (0)(-1) = 0$$
$$\Sigma c_{1} c_{3} = (1)(\tfrac{1}{2}) + (-1)(\tfrac{1}{2}) + (0)(-\tfrac{1}{2}) + (0)(-\tfrac{1}{2}) = 0$$
$$\Sigma c_{2} c_{3} = (0)(\tfrac{1}{2}) + (0)(\tfrac{1}{2}) + (1)(-\tfrac{1}{2}) + (-1)(-\tfrac{1}{2}) = 0$$

When all possible pairs of comparisons are found to be orthogonal, they are said to be an orthogonal set.

The results of a set of orthogonal linear comparisons can be presented in an ANOVA table as in Table 7.12. The tabular F value $[F_{0.05(1,20)}]$ is 4.35. Therefore, the null hypothesis of no difference is accepted for the first and second comparison. However, for the third comparison, the null hypothesis would be rejected and the researcher would conclude that steers fed a protein gained more than steers fed a nonprotein nitrogen.

TABLE 7.12.   ORTHOGONAL LINEAR COMPARISONS FOR THE BEEF STEER FEEDING TRIAL

Source of Variation	df	SS	MS	F
Soybean vs cottonseed	1	10	10	<1
Urea vs ammonia	1	102	102	1.74
Protein vs nonprotein	1	581	581	9.92
Within diets	20	1171	58.55	
Total	23	1863		

Orthogonal Polynomial Comparisons

As a second example of this type of comparison, consider a special kind of linear comparisons known as orthogonal polynomial comparisons. These comparisons specifically test a set of quantitatively spaced treatments to determine if the response to the treatment levels is linear, quadratic, cubic, etc. To illustrate this procedure suppose that the dietary treatments are increasing levels of urea in the diet.

Treatment	Mean
1—0% Urea	91.67
2—0.5% Urea	89.83
3—1.0% Urea	83.83
4—1.5% Urea	78.00

The constants for the orthogonal linear comparisons for equally spaced quantitative treatment levels are found in Appendix Table F.1. For 4 treatments the contrasts are as given:

Treatment	1	2	3	4
Linear (L)	−3	−1	+1	+3
Quadratic (Q)	+1	−1	−1	+1
Cubic (C)	−1	+3	−3	+1

$$SS_L = \frac{6[(-3)(91.67) + (-1)(89.83) + (1)(83.83) + (3)(78.00)]^2}{(-3)^2 + (-1)^2 + (1)^2 + (3)^2}$$

$$= 663.0$$

$$SS_Q = 23.9$$

$$SS_C = 5.6$$

If you complete the ANOVA, you will find that only the linear comparison is significant. Therefore we would conclude that gain decreases linearly with increasing levels (0 to 1.5%) of urea in the diet.

The second comparison procedure is the least significant difference (lsd). This is a paired comparison and therefore the Type I error rate for the experiment (total probability of a Type I error in the experiment) can be very large. For example, if all paired comparisons are carried out at the 5% level, with 4 treatments, the probability of a Type I error in the experiment is $\cong 0.21$. However, the use of lsd has been fairly common, primarily due to its simplicity of application.

$$\text{lsd} = t_{\alpha(df_w)}S_{\bar{D}}$$

where t_{α} is the tabular Student's t value with α probability and degrees of freedom within and $S_{\bar{D}}$ is the standard error of the difference computed as:

$$S_{\bar{D}} = \left(\frac{2MS_W}{n}\right)^{1/2}$$

For the feeding trial data, the 5% lsd equals:

$$\text{lsd} = t_{0.05(20)}\left(\frac{2MS_W}{n}\right)^{1/2}$$

$$= 2.086\left(\frac{2(58.55)}{6}\right)^{1/2}$$

$$= 9.2$$

The interpretation is that a difference between any two treatment means which exceeds the lsd value is significant at the α level. Therefore, the means might be presented as follows where means with a common line are not significantly different.

Treatment	1	2	3	4
Mean	91.7	89.8	83.8	78.0

In general, because of the high experimental error rate (Type I), lsd should not be used to make all possible comparisons. It is recommended that its use be limited to not more than a -1 pre-planned comparisons.

Duncan's New Multiple Range Test

The third method of comparison is Duncan's new multiple range test. This procedure was specifically developed to make all possible paired comparisons between treatment means, of which there are a(a − 1)/2 comparisons. The concept was to develop a technique for which the experimental Type I error rate was approximately the same as that for a set of a − 1 orthogonal comparisons.

The procedure is as follows:
Find the standard error of the treatments

$$S_{\bar{Y}} = \left(\frac{MS_W}{n} \right)^{1/2}$$

$$= \left(\frac{58.55}{6} \right)^{1/2} = 3.12$$

From Appendix Table G.1 obtain the significant studentized range (SSR) values for df_W and p. Where p = 2 to a, corresponds to the number of means in the range of the 2 means being compared when the means are ranked in order of magnitude. Then compute the least significant range (LSR) as

$$LSR = SSR(S_{\bar{Y}})$$

For our data set we find

P	2	3	4
$SSR_{0.05}$	2.95	3.10	3.18
LSR	9.2	9.7	9.9

Now rank the treatment means in order of magnitude:

Treatment	4	3	2	1
Mean	91.7	89.8	83.8	78.0

Then proceed to test the means, largest to smallest, largest to second smallest, . . ., second smallest to smallest. The first comparison would be $T_4 - T_1 = 13.7 > 9.9$; $T_4 - T_2 = 7.9 < 9.7$,. . .; $T_2 - T_1 = 5.8 < 9.2$. Each difference is declared significant at the α level if it is greater than the corresponding LSR value. To prevent apparent inconsistencies in the results, a rule is also applied that states that no difference may be declared significant if the means being compared lie in the range of a pair of means found non-significantly different. In this case, our results would be the same as for the lsd, although in general we would expect to find fewer significant differences, while the Type I error rate would be smaller.

When publishing results, presentation of means in rank order is likely to be inconvenient. A common method of presenting results is the superscripting procedure. For example:

Treatment	Mean
1	78.0^b
2	$83.8^{a,b}$
3	89.8^a
4	91.7^a

A footnote is included which states that means with any like superscripts are not significantly different at the 5% level by Duncan's multiple range test. This format allows the writer to present the means in any order so long as the superscripts are carried with the mean. Note that a line in the underlining procedure is assigned a letter in the superscript procedure.

If the reader would like more information concerning mean comparison procedures, Steel and Torrie (1960) discuss several comparison procedures and Chew (1977) discusses the problems of sensitivity and error rate for a number of comparison procedures.

PROBLEMS

7.1. To study the 1-way analysis of variance without getting lost in a large data set consider the following simple data set:
Group 1—22, 15, 29
Group 2—36, 51, 42

(a) Compute the F ratio directly from the data using the sample variances and the variance of the means.
(b) Complete the ANOVA table using the formulas for the computing SS.
(c) Given the model $Y_{ij} = \mu + \alpha_i + \epsilon_{ij}$, find the sample estimates for the 3 terms in the model for each observation.

7.2. A horticulturist conducted an experiment to study the effectiveness of 4 chemicals in reducing plant size to produce dwarf plants. The experiment included a nontreated control and the 4 chemical treatments with 7 plants per treatment. The experiment was conducted in a greenhouse in a completely random arrangement with a single plant in each pot. The measured variable was node length (cm) and was recorded as follows:

Pot	None	A	B	C	D
			Chemical		
1	9.9	8.5	4.0	10.0	9.5
2	9.6	7.4	5.8	7.7	11.2
3	10.5	7.5	5.1	8.1	10.3
4	9.9	8.5	5.8	9.8	9.3
5	8.0	9.5	5.7	11.0	7.6
6	8.9	8.5	4.6	9.8	8.3
7	11.5	10.4	6.2	8.1	9.9

(a) Complete the 1-way ANOVA.
(b) Apply the Duncan's multiple range test to determine which treatment groups are different.
(c) Suppose that chemicals A and B, and that chemicals C and D represent X and Y at low and high dosage levels, respectively.

Construct a set of orthogonal comparisons, and partition the treatment SS into 5 single degrees of freedom comparisons.

7.3. A chemical laboratory is concerned with the variability among and within their technicians to perform certain laboratory analyses. A single source of material was obtained and 4 randomly selected technicians were asked to complete a certain analysis several times. The following data (%) were obtained:

Determination	Technician 1	Technician 2	Technician 3	Technician 4
1	3.72	3.51	3.67	3.49
2	3.69	3.63	3.68	3.52
3	3.43	3.72	3.72	3.55
4	3.91	3.41	3.59	
5	3.73		3.64	
6	3.42			

(a) Complete the 1-way ANOVA.
(b) Assuming that the data represent a random model, determine the variance components for among and within technicians.
(c) If the laboratory were asked to conduct a single determination on an unknown sample, what would be your best estimate of the variance associated with the determination?

7.4. Conduct an analysis of variance for the data of problem 5.1. Note that the $t^2 = F$, since ANOVA and t-test are equal for tests with 1 degree of freedom in numerator.

REFERENCES

For a nontheoretical presentation of the basic concepts of ANOVA see:
SOKAL, R.R. and ROHLF, F.J. 1969. Biometry. W.H. Freeman and Company, San Francisco.
STEEL, R.G.D. and TORRIE, J.H. 1976. Introduction to Statistics. McGraw-Hill Book Company, New York.

For additional examples of ANOVA applications see:
OSTLE, B. and MENSING, R.W. 1975. Statistics in Research. Iowa State Univ. Press, Ames.
ZAR, J.H. 1974. Biostatistical Analysis. Prentice-Hall, Englewood Cliffs, N.J.

For mean comparison procedures see:
CHEW, V. 1977. Comparisons among treatment means in an analysis of variance. U.S. Dep. Agric., Agric. Res. Serv., *ARS-H16*.

STEEL, R.G.D. and TORRIE, J.H.   1960.   Principles and Procedures of Statistics.   McGraw-Hill Book Company, New York.

A more theoretical approach to ANOVA may be found in:

GUENTHER, W.C.   1964.   Analysis of Variance. Prentice-Hall, Englewood Cliffs, N.J.

SCHETTE, H.   1959.   The Analysis of Variance.   John Wiley & Sons, New York.

Examples of 1-way ANOVA, multiple range test, and linear contrast applications in recent literature:

FAHEY, G.C., JR., FRANK, G.R., JENSEN, A.H. and MASTERS, S.S.   1980. Influence of various purified and isolated cell wall fibers on the utilization of certain nutrients by swine and hamsters.   J. Food Sci. *45*, 1675.

LEE, K. and CLYDESDALE, F.M.   1980.   Effect of baking on the forms of iron in iron-enriched flour.   J. Food Sci. *45*, 1500.

LESING, G., KLOPFENSTEIN, T., RUSH, I. and WARD, J.   1980.   Chemical treatment of wheat straw.   J. Anim. Sci. *51*, 263.

McMILLIN, K.W., SEBRANEK, J.G., RUST, R.E. and TOPEL, D.G.   1980. Chemical and physical characteristics of frankfurters prepared with mechanically processed pork product.   J. Food Sci. *45*, 1455.

MILLER, A.J., ACKERMAN, S.A. and PALUMBO, S.A.   1980.   Effects of frozen storage on functionality of meat for processing.   J. Food Sci. *45*, 1466.

8

Analysis of Variance II: The Multiway ANOVA

In the previous chapter we dealt with data for which only one assignable source of variation exists (CRD). All other sources of variation were part of the within-group variation which was collectively referred to as experimental error. In this chapter, we will consider designs in which more than one assignable source of variation is computed. Hence, analysis for these designs is referred to as "multiway ANOVA."

In the completely random design, the assignable source of variation was the treatment (or group) effect. In multiway analysis, the additional assignable sources of variation are used to account for expected lack of uniformity in the experimental units independent of treatment effects. If these additional sources were left unaccounted for, the result would be a loss in the sensitivity (ability to detect real treatment effects) since the within treatment source of variation would be inflated by the additional "noise" associated with the heterogeneity among the experimental units.

For example, suppose that due to physical limitations a researcher finds that he is not able to conduct all of his experiment at the same time. Therefore, he must conduct the experiment during more than one time period. If the researcher properly designs his study, time effects will become an assignable source of variation and, as such, will not increase his estimate of experimental variation.

The two multiway classifications that will be considered in this chapter are the randomized complete block (RCB) and the Latin square designs. Although these two designs are the two more common multiway classifications, they are not the only ones available to the researcher. The reader wanting additional information about multiway designs should consult a design text such as Cochran and Cox (1957).

THE LINEAR ADDITIVE MODEL (2-WAY)

Just as for the 1-way model, it is helpful for the reader to examine the statistical model for the 2-way analysis. It is common to refer to the addi-

108

tional source of variation in the 2-way model as the "block effect." The statistical model for the 2-way analysis of variance again assumes a linear model in which the effects are additive. As with the 1-way model, it is assumed that the experimental errors are normally and independently distributed with equal variances.

Symbolically, we would present the model for the 2-way ANOVA as:

$$Y_{ij} = \mu + \alpha_i + \beta_j + \epsilon_{ij}$$

where $i = 1,\ldots,a$

$j = 1,\ldots,n$

μ is the true mean

α_i is the true deviation of the mean of the ith treatment (group) from μ

β_j is the true deviation of the mean of the jth block from μ

and ϵ_{ij} is the random deviation of the observation of the jth block and the ith treatment from the true mean of the ijth cell ($\mu_{ij} = \mu + \alpha_i + \beta_j$).

These deviations are assumed to be normally and independently distributed with a mean of zero and a common variance [NID $(0,\sigma^2)$].

Again, a description of what the linear additive model says is in order. The value of each observation (Y_{ij}) is due to 4 effects. The first is the grand mean (μ), the second is how much the treatment mean (μ_i) deviates from the grand mean, the third is how much the block mean (β_j) deviates from the grand mean, and the fourth is how much the observation Y_{ij} deviates from the mean of the ijth treatment-block combination (μ_{ij}). Estimates of these effects may be computed as:

$$\hat{\mu} = \bar{Y}_{..}$$
$$\hat{\alpha}_i = (\bar{Y}_{i.} - \bar{Y}_{..})$$
$$\hat{\beta}_j = (\bar{Y}_{.j} - \bar{Y}_{..})$$
$$\hat{\epsilon}_{ij} = (Y_{ij} - \bar{Y}_{i.} - \bar{Y}_{.j} + \bar{Y}_{..})$$

Although we have again identified the ϵ_{ij} as a random effect, our description of the model is not complete without further classification of the α_i and β_j effects. The researcher may choose treatments and/or blocks such that their effects are either fixed or random. This concept of an effect being fixed or random is the same as that presented for the 1-way classification. In application, the most common situation is for both α_i and β_j to be regarded as fixed ($\Sigma\alpha_i = 0$ and $\Sigma\beta_j = 0$).

When both block and treatment effects are regarded as fixed, then an additional assumption is required for an exact test of treatment effects. The assumption is that the treatment effects are the same in each block, that is, the relative performance of the treatments does not vary from block to block. In other words, there is no "interaction" between blocks and treatments. When block and treatment interactions are present, it can be shown that the interaction effect is present in the estimate of experimental error, but absent in the mean square for treatment effects. Although this assumption is necessary for the simple model described here, in the next chapter we will consider models which include tests for interaction effects. If in practice such interactions are likely to occur, then the researcher should design his experiment such that the analysis will include the interaction as an assignable source of variation.

If the researcher should decide that either the block or treatment effects may be regarded as random, then a different situation prevails. Suppose that the blocks were randomly selected from a population of possible blocks. In this case the block effects are assumed to be [NID $(0,\sigma_\beta^2)$]. Since the blocks are selected at random, it is assumed that the researcher is, in this case, interested in making generalizations about the treatment effects on the population of blocks. Note that in the fixed case the statistical conclusions apply specifically to the selected blocks.

In the case of random blocks and fixed treatments it is not necessary to assume no interaction of block and treatment effects for an exact test of treatment effects. This is true because the interaction is present in both the mean square for experimental error and the mean square for treatment effects leading to an expected F ratio of 1 in the absence of treatment effects. However, the presence of interaction effects still leaves the researcher in a difficult position. If experimental units do not respond to treatments the same way in different blocks (interaction), general conclusions about treatment effects are likely to be of little value. A further discussion of the problem is presented in the following chapter.

Since in the simple multiway ANOVA models, interaction effects cannot be estimated, these models should not be used when interaction might reasonably occur. In such a case, models including interaction effects presented in the following chapter will better suit the experimental situation.

THE RANDOMIZED COMPLETE BLOCK DESIGN

As pointed out in the previous chapter, the completely random design is ideally suited for experimental units that are essentially homogeneous. If the experimental material is relatively homogeneous, the use of a design more complex than the completely random design would not be advantageous. However, in many cases the researcher will not have available for research a set of homogeneous experimental units. It is in this latter case that more complex models are of considerable use.

The purpose of the randomized complete block (RCB) design is then to account for an additional source of variation, other than treatment (inherent or induced by experimental methods) in the experimental material. The concept of the RCB design is to group the experimental material into groups called "blocks" such that the differences between units are as little as possible within blocks while the differences between blocks are large. In the simple RCB design which will be considered here, the number of experimental units in each block is equal to the number of treatments. Treatments are then assigned at random within each block of the experiment such that each treatment occurs once and only once. This simple RCB design is the only one that will be considered in this text. The reader should keep in mind that a number of variations exist which permit greater numbers of units per block than treatments.

This arrangement in effect results in each block being a single replicate of the experiment. Since blocks have been constructed to maximize within block uniformity, differences between units within a block should for the most part be due to treatment effects. Also since every block contains each treatment the same number of times (once for the simple RCB design), differences between blocks are due to heterogeneity among experimental units and not treatment effects. Thus the block differences become an assignable source of variation and therefore account for a portion of variability that would otherwise have been included in the estimate of experimental error.

The control of experimental error is then the major function of multiway designs. As referred to in Chapter 6, this is known as the principle of "local control." The researcher who chooses to use a blocking design is simply assigning an identifiable portion of the variation to blocks, while the experimental error represents variation over which the researcher has no control. It is this remaining variation which we assume to be NID $(0, \sigma^2)$.

Now let us consider some examples which illustrate the concept of blocking. Probably the most common application of the RCB design is in field plot research. Variation in field plots is frequently rather marked even over a small area. Probably the major reason is soil fertility, although soil type, insect infestation, disease, and topography, etc., play a role. The grouping of plots into blocks, even in the absence of specific knowledge about the plots, is not a major problem in field plot research. Since adjacent plots are more likely to be similar than plots located apart, creating blocks of adjacent plots is generally successful. In cases where additional information is available (i.e., plot fertility) other blocking arrangements may be better.

A similar blocking procedure is frequently valuable in studies conducted in greenhouses, growth chambers, and animal studies. That is, blocking on location (or position) will frequently be of value. Even in the controlled conditions found in the greenhouse or growth chamber, variation in light intensity, watering patterns, temperature, and/or air movement, results in adjacently located experimental units being more alike than units in separated areas. In animal studies, similar environmental conditions may indicate natural blocking arrangements. For example, rats located on different tiers of a rat rack may form a block. In large animal studies, pens, all located on one side of the barn, may be a natural block.

The replication of an experiment in time is probably another good example of a natural blocking concept. If a complete set of treatments is run in each of several time periods, then time periods may be thought of as blocks. Experimental units in one time period are likely to be similar, while differences between blocks (time periods) should not be influenced by treatments. The time periods may be relatively short, such as minutes, hours, or days in the case of laboratory experiments, or as long as years in the case of field studies.

Before we move on to the analysis of the RCB experiment, some comments need to be made regarding the conduct of RCB experiments. We have discussed the process of blocking and the random assignment of treatments within blocks in the chapter on experimental design. In general, throughout the experiment, the processing of experimental units should be done keeping in mind the concept of maintaining maximum uniformity among units within blocks. If changes in procedures, methods, materials, etc., are necessary during the study, they may be made between blocks. Any procedure or process within blocks which might result in differences between units should be carried out in random order. The order of processing blocks may be systematic when estimates of the block effects are not of interest to the researcher.

ANOVA—SIMPLE RCB DESIGN

To illustrate the simple RCB design, suppose that an agronomist wishes to conduct a variety trial to compare the yield of 6 varieties of soybeans with 4 blocks. An example of the field plot layout is given in Fig. 8.1. The 6 treatments (varieties) have been randomly assigned in each of the 4 blocks. When the soybeans are harvested, yield data are recorded on each plot.

Even though plot size was probably considerably less than 1 acre (0.4 ha), yields are reported on a bu per acre (67.2 kg/ha) basis (Table 8.1). The

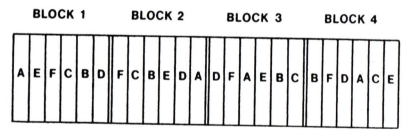

FIG. 8.1.   FIELD PLOT LAYOUT (RCB DESIGN) FOR DATA OF TABLE 8.1

TABLE 8.1.   SOYBEAN YIELDS (BU/ACRE) FOR 6 VARIETIES IN AN RCB DESIGN

| Block | Variety | | | | | | |
	A	B	C	D	E	F	$Y_{.j}$
1	42	42	41	25	31	36	217
2	33	37	33	27	16	12	158
3	31	37	33	35	29	11	176
4	21	19	14	13	17	13	97
$Y_{i.}$	127	135	121	100	93	72	$Y_{..} = 648$

1 bu/acre = 67.2 kg/ha.

process of changing data from one kind of measurement unit to another measurement unit does not in any way alter the significance of the data. Note that the only new summations needed for the RCB design are the block sums.

The sum of squares equations are presented in Table 8.2. Again, both definitional and working formulas are given to help the reader better understand the source of variation being computed. Using the computing equations, the sums of squares for the RCB are computed as follows:

$$SS_{Blocks} = \frac{217^2 + 158^2 + 176^2 + 97^2}{6} - \frac{648^2}{(6)(4)}$$

$$= 1243.7$$

$$SS_{Varieties} = \frac{127^2 + 135^2 + \ldots + 72^2}{4} - \frac{648^2}{(6)(4)}$$

$$= 711.0$$

$$SS_{Total} = (42^2 + 33^2 + \ldots + 13^2) - \frac{648^2}{(6)(4)}$$

$$= 2502.0$$

$$SS_{Error} = SS_{Total} - SS_{Blocks} - SS_{Varieties}$$

$$= 547.3$$

TABLE 8.2.   DEFINITIONAL AND WORKING SUMS OF SQUARES FORMULAS FOR THE RCB ANOVA

Source of Variation	df	Sums of Squares	
		Definition	Computing
Block	$b - 1$	$a\Sigma(\bar{Y}_j - \bar{Y}..)^2$	$\dfrac{\Sigma Y^2_{.j}}{a} - \dfrac{Y^2..}{ab}$
Treatment (varieties)	$a - 1$	$b\Sigma(\bar{Y}_{i.} - \bar{Y}..)^2$	$\dfrac{\Sigma Y^2_{i.}}{b} - \dfrac{Y^2..}{ab}$
Experimental error	$(b-1)(a-1)$	$\Sigma(Y_{ij} - \bar{Y}_j - \bar{Y}_{i.} + \bar{Y}..)^2$	Subtraction
Total	$ab-1$	$\Sigma(Y_{ij} - \bar{Y}..)^2$	$\Sigma Y^2_{ij} - \dfrac{Y^2..}{ab}$

The complete ANOVA is presented in Table 8.3. The F ratio for varieties is a test of the null hypothesis that the variety means are all equal. Since the F ratio is computed as MS_V/MS_E, we have 5 and 15 degrees of freedom for the numerator and denominator, respectively. By examination of Appendix

Table D.1, we find that $F_{0.05(5,15)}$ = 2.90. Therefore, since the sample F is greater than the tabular F (3.90>2.90), we would reject the null hypothesis (P<0.05) and conclude that differences in yield are likely for the varieties in the experiment. If the researcher wished to know which varieties were different, one of the mean comparison procedures presented in Chapter 7 would be applied.

TABLE 8.3.   COMPLETE ANOVA FOR THE SOYBEAN VARIETY TRIAL (RCB)

Source of Variation	df	SS	MS	F
Block	3	1243.7	414.6	—
Variety	5	711.0	142.2	3.90
Error	15	547.3	36.5	—
Total	23	2502.0	—	—

Having completed the RCB ANOVA it is of some interest to examine the gain (or loss) in efficiency by using the RCB design. The relative efficiency of the CRD to the RCB design may be computed from information in the RCB ANOVA as follows:

$$\text{R.E.} = \frac{(b - 1)(MS_B) + b(a - 1)MS_E}{(ab - 1)MS_E}$$

For the data of Table 8.3 we compute relative efficiency as:

$$\text{R.E.} = \frac{3(414.6) + 20(36.5)}{23(36.5)}$$

$$= 2.35$$

Relatively efficiency is an estimate of the ratio of the 2 error mean squares for alternative designs. That is, R.E. is an estimate of MS_{CRD}/MS_{RCB}.

Although the comparison is not perfect, since the degrees of freedom are not identical for the 2 mean squares, the comparison is generally adequate except when the degrees of freedom for the error MS_{RCB} are quite small. The interpretation of the R.E. measurement is that, for our example, approximately 2.35 times as many replications would be needed using similar experimental units if a CRD were used.

EXPECTED MEAN SQUARES FOR RCB

In the previous chapter we discussed variance components and F ratios for the 1-way ANOVA. Again it is useful to examine what the average or expected value is for each of the mean squares in the RCB ANOVA. The expected MS's are presented in Table 8.4 under the assumption of no interaction of block and treatment effects. In Chapter 9 we will discuss in some detail the concept of interactions. Further discussion of the effect of interactions on the expected mean squares for the RCB ANOVA will be considered at that time.

For each model it should be apparent that the appropriate F statistic for testing the null hypothesis about treatment effects is $F = \dfrac{MS_T}{MS_E}$. For example, consider the fixed model. The expected F ratio would equal

$\dfrac{\sigma^2 + \dfrac{b}{a-1}\,\Sigma\alpha_i^2}{\sigma^2}$. If the treatment effects are absent, then the $\Sigma\alpha_i^2 = 0$ and the F ratio would have an expected value of 1 for a true null hypothesis $F = \dfrac{\sigma^2 + 0}{\sigma^2} = 1$. The same relationship holds for the random and mixed models.

TABLE 8.4.   EXPECTED MEAN SQUARES FOR THE SIMPLE RCB DESIGN

		Expected Mean Squares	
Source	Fixed Model	Random Model	Mixed Model
Block	$\sigma^2 + \dfrac{a}{b-1}\,\Sigma\beta_j^2$	$\sigma^2 + a\sigma^2_\beta$	$\sigma^2 + a\sigma^2_\beta$
Treatment	$\sigma^2 + \dfrac{b}{a-1}\,\Sigma\alpha_i^2$	$\sigma^2 + b\sigma^2_\alpha$	$\sigma^2 + \dfrac{b}{a-1}\,\Sigma\alpha_i^2$
Error	σ^2	σ^2	σ^2

MISSING DATA

Sometimes during the course of an experiment, individual experimental units are lost or the data from an individual experimental unit become unusable. To maintain the RCB structure of the data, we could simply drop that block or treatment from the data set and proceed with the RCB analysis. The discarding of data is generally unappealing and the remaining data in the block or treatment do provide some useful information. Techniques have been developed by Yates to estimate a value for the missing experimental unit. The estimated value cannot in any way be viewed as an experimental observation. Its sole purpose is to facilitate the RCB analysis, thus salvaging the remaining data in the block and treatment where the missing observation is located.

The following equation may be used to estimate a single missing value:

$$Y_{ij} = \frac{bY_{\cdot j} + aY_{i\cdot} - Y_{\cdot\cdot}}{(b-1)(a-1)}$$

where $Y_{\cdot j}$, $Y_{i\cdot}$, and $Y_{\cdot\cdot}$ are the totals of the remaining observations for the corresponding block and treatment.

Suppose, for example, that the observation for block 2, variety D was unavoidably lost. The remaining data would appear as in Table 8.5.

$$Y_{42} = \frac{4(131) + 6(73) - 621}{(3)(5)}$$

$$= 22.7$$

TABLE 8.5.   EXAMPLE FOR MISSING DATA ESTIMATION

| Block | Variety | | | | | | |
	A	B	C	D	E	F	Y_{ij}
1	42	42	41	25	31	36	217
2	33	37	33	( )	16	12	(131)
3	31	37	33	35	29	11	176
4	21	19	14	13	17	13	97
$Y_{i\cdot}$	127	135	121	(73)	93	72	$Y.. = (621)$

Had the yield for block 2, variety D been missing, we would have replaced it with a value of 22.7 for purposes of calculation. After replacement, $Y_{4\cdot} = 95.7$, $Y_{\cdot2} = 153.7$, and $Y.. = 643.7$.

The sums of squares for the analysis of variance would be computed using the above sums and the estimated value $Y_{42} = 22.7$. The revised analysis of variance is presented in Table 8.6. Note that the error and total degrees of freedom have been reduced by 1 since an estimated value in no way supplies any real information. The equation for the missing value is based on providing a minimum variance for the experimental error; however, the treatment mean square is biased upwards by $\dfrac{[Y_{\cdot j}-(a-1)Y_{ij}]^2}{a(a-1)^2}$ where $Y_{\cdot j}$ and Y_{ij} are the sum of the remaining values in the block and the estimate of the missing value, respectively.

TABLE 8.6.   ANOVA FOR RCB DESIGN WITH A MISSING OBSERVATION

Source	df	SS	MS	F
Block	3	1251.7	—	—
Variety	5	732.1	135.6	3.54
Error	14	535.9	38.3	—
Total	22	2519.7	—	—

The adjusted varieties MS equals:

$$MS_{V\ adj} = 146.4 - \frac{[153.7-(5)(22.7)]^2}{(6)(5^2)}$$

$$= 135.6$$

The standard error of the mean including the estimate of the missing value is

$$S_{\bar{Y}} = \left(MS_E \left[\frac{1}{b} + \frac{a}{b(b-1)(a-1)} \right] \right)^{1/2}$$

When several values are missing, the analysis of the data becomes more difficult. At this point it is probably desirable to consult a statistician or to consult a more advanced text concerning the subject. Methods for estimating multiple missing values have been suggested by Yates. When more than one missing value occurs in the same treatment or block, the researcher should give serious consideration to deleting the treatment or block from the analysis.

THE LATIN SQUARE DESIGN

The Latin square design makes use of the blocking or grouping concept introduced with the randomized complete block design. While the randomized complete block design deals with one blocking factor, the Latin square may be thought of as a double blocking arrangement. Although we commonly refer to the two blocking factors in the Latin square design as row and column effects, these are groups of experimental units arranged to permit the measurement of two identifiable sources of variation plus treatment effects.

A Latin square design with 4 treatments is presented in Fig. 8.2. Assuming that the 4 treatments are identified by letters, note that each treatment occurs exactly once in each row and column. This arrangement of treatments results in each row and each column being a simple complete block. The appropriate analysis of variance will have assignable sources of varia-

Columns

	1	2	3	4
I	D	C	A	B
II	A	D	B	C
III	B	A	C	D
IV	C	B	D	A

Rows

FIG. 8.2. FIELD PLOT LAYOUT FOR LATIN SQUARE DESIGN

tion corresponding to row, column, and treatment effects, as well as the estimate of the random experimental error.

The statistical model for the Latin square design assumes that the effects are additive and that the experimental errors are normally and independently distributed with equal variances. It is further assumed that interactions do not exist among row, column, or treatment effects. The linear additive model would be presented as:

$$Y_{ijk} = \mu + \alpha_i + \beta_j + \gamma_k + \epsilon_{ijk}$$

where i   = 1,. . .,a
      j   = 1,. . .,a
      k   = 1,. . .,a
      μ is the true mean
      α_i is the true deviation of the mean of the ith treatment from μ
      β_j is the true deviation of the mean of the jth row from μ
      γ_k is the true deviation of the mean of the kth column from μ
and ϵ_{ijk} are random deviations assumed to be NID $(0,\sigma^2)$.

Several things about the model deserve further mention. Although each observation carries 3 subscripts, the presentation of the data is 2-dimensional (Fig. 8.2). If 4 treatments are to be evaluated, then 4 rows and 4 columns are required. Note, however, that there are not $4 \times 4 \times 4 = 64$ observations, but rather only 16. In this case, 48 of the possible combinations are not included in the study. Experiments which include all combinations will be considered in the following chapter. It is this partial set of combinations which makes the assumption of no interactions necessary. When a researcher finds the assumption of no interaction untenable, then the Latin square should not be used.

Again, a complete characterization of the effects requires that each be identified as fixed or random. Therefore various sets of expected mean squares corresponding to the fixed, random, or a number of mixed models could be present. Table 8.7 presents the fixed and random model. For the

TABLE 8.7.   EXPECTED MEAN SQUARES FOR THE LATIN SQUARE DESIGN

Source	Expected Mean Squares	
	Fixed Model	Random Model
Row	$\sigma^2 + a\dfrac{\Sigma\beta_j^2}{a-1}$	$\sigma^2 + a\beta^2$
Column	$\sigma^2 + a\dfrac{\Sigma\gamma_k^2}{a-1}$	$\sigma^2 + a\sigma_\gamma^2$
Treatment	$\sigma^2 + a\dfrac{\Sigma\alpha_i^2}{a-1}$	$\sigma^2 + a\sigma_\alpha^2$
Error	σ^2	σ^2

mixed model, the random component would be substituted for the corresponding fixed component for random effects. Regardless of the model, the F ratio for treatments (also for rows or columns) uses the MS_E for the denominator.

Now let us turn our attention to the use of the Latin square design in an experiment. The 4 × 4 Latin square presented in the first of this section is an example of a randomized Latin square. To obtain a randomized Latin square the following steps are necessary.

First write down any Latin square of the dimension desired. It may be systematically written as in this example, or "starter" squares may be obtained in a number of design texts.

	(1)	(2)	(3)	(4)
(1)	A	B	C	D
(2)	D	A	B	C
(3)	C	D	A	B
(4)	B	C	D	A

Now obtain from a random number table a random sequence of the numbers 1 through 4. Say we draw the sequence 2, 1, 3, 4. Now rearrange the Latin square by column in the random sequence.

	(2)	(1)	(3)	(4)
(1)	B	A	C	D
(2)	A	D	B	C
(3)	D	C	A	B
(4)	C	B	D	A

Now follow the same procedure to rearrange the rows. Suppose the random sequence is 3, 2, 1, 4. We now obtain the 4 × 4 Latin square presented earlier.

	(2)	(1)	(3)	(4)
(3)	D	C	A	B
(2)	A	D	B	C
(1)	B	A	C	D
(4)	C	B	D	A

The final step is to randomly assign the actual treatments to the letters A, B, C, and D.

Consider the following example which shows how the ANOVA for the Latin square is conducted. Suppose that the researcher is studying the effect of different levels of a certain atmospheric pollutant on plant growth. The study consists of 4 treatments and only 4 growth chambers are available. Replication of the study will be over time. In the Latin square arrangement, let columns correspond to growth chambers, rows to replication in time, and the letters to the 4 treatments. In Table 8.8, we have the dry matter weights of the plants resulting from the 4 × 4 Latin square design.

TABLE 8.8.   DRY MATTER WEIGHT IN G FOR PLANTS SUBJECTED TO DIFFERENT CONCENTRATIONS OF A POLLUTANT IN A 4 × 4 LATIN SQUARE DESIGN

		Growth Chamber (k)				
		1	2	3	4	$Y_{.j.}$
Time period (j)	I	D 345	C 345	A 299	B 363	1352
	II	A 355	D 376	B 319	C 407	1457
	III	B 384	A 375	C 298	D 358	1415
	IV	C 368	B 405	D 364	A 439	1576
	$Y_{..k}$	1452	1501	1280	1567	$Y_{...} = 5800$
		Treatment (i) Totals				
		A	B	C	D	
	$Y_{i..}$	1468	1471	1418	1443	

In Table 8.9, the definitional and computing sums of squares formulas are given. Definitional formulas are based on means to provide a better understanding of the effect being tested. The computing formulas were used to compute the following sums of squares:

$$SS_T = \frac{1468^2 + 1471^2 + 1418^2 + 1443^2}{4} - \frac{5800^2}{16}$$

$$= 460$$

$$SS_R = \frac{1352^2 + 1457^2 + 1415^2 + 1576^2}{4} - \frac{5800^2}{16}$$

$$= 6689$$

$$SS_C = \frac{1452^2 + 1501^2 + 1280^2 + 1567^2}{4} - \frac{5800^2}{16}$$

$$= 11299$$

$$SS_{Total} = 345^2 + 355^2 + \ldots + 439^2 - \frac{5800^2}{16}$$

$$= 21246$$

$$SS_{Error} = SS_{Total} - SS_T - SS_R - SS_C$$

$$= 21246 - 460 - 6689 - 11299$$

$$= 2798$$

TABLE 8.9. DEFINITIONAL AND COMPUTING FORMULAS FOR SUMS OF SQUARES FOR THE LATIN SQUARE ANOVA

Source	df	Definitional	Working
			Sums of Squares
Row	$a - 1$	$a\Sigma (\bar{Y}_{j\cdot} - \bar{Y}\ldots)^2$	$\dfrac{\Sigma Y^2_{\cdot j\cdot}}{a} - \dfrac{Y^2\ldots}{a^2}$
Column	$a-1$	$a\Sigma (\bar{Y}_{\cdot\cdot k} - \bar{Y}\ldots)^2$	$\dfrac{\Sigma Y^2_{\cdot\cdot k}}{a} - \dfrac{Y^2\ldots}{a^2}$
Treatment	$a-1$	$a\Sigma(\bar{Y}_{i\cdot\cdot} - \bar{Y}\ldots)^2$	$\dfrac{\Sigma Y^2_{i\cdot\cdot}}{a} - \dfrac{Y^2\ldots}{a^2}$
Error	$(a-1)(a-2)$	$\Sigma(Y_{ijk} - \bar{Y}_{j\cdot} - \bar{Y}_{\cdot\cdot k} - \bar{Y}_{i\cdot\cdot} + 2\bar{Y}\ldots)^2$	Subtraction
Total	a^2-1	$\Sigma(Y_{ijk} - \bar{Y}\ldots)^2$	$\Sigma Y^2_{ijk} - \dfrac{Y^2\ldots}{a^2}$

Table 8.10 is the complete ANOVA for the Latin square. The null hypothesis being tested is that the treatment means are equal. Since the F ratio for treatments is much less than 1, intuitively we should suspect that the null hypothesis cannot be rejected. If we examine the F table we find $F_{0.05(3,6)} = 4.76$, which is much greater than the sample F ratio of 0.33. Therefore we conclude that the concentrations of the pollutant used in this study did not affect plant growth (dry matter weight).

TABLE 8.10. COMPLETE ANOVA FOR GROWTH CHAMBER DATA OF TABLE 8.8 (4 × 4 LATIN SQUARE)

Source of Variation	df	SS	MS	F
Row (time)	3	6689	2229.7	—
Column (chamber)	3	11299	3766.3	—
Treatment	3	460	153.3	0.33
Error	6	2798	466.3	—
Total	15	21246	—	—

Upon completion of the Latin square the researcher may, as in the case of the RCB, examine the relative efficiency of the Latin square design relative to the RCB or CRD. Equations for these comparisons may be found in most basic design texts. Likewise, equations for estimating missing values and adjustment treatment sums of squares are also, if needed, available in design texts.

THE SIZE OF THE EXPERIMENT

One of the more common questions asked of a statistician is how many replications are necessary for a given test of significance. Unfortunately the answer to this question is not easy. Without an estimate of the experimental variability, no solution is possible. Estimates of variability may be obtained from analysis of previous experiments such as those in previous sections. Let us examine a procedure which will answer the following question. How many replicates (r) will be needed to find a given true difference (δ) between any 2 treatment means significant, at the α level, with a probability P (P is the power of the test), if δ exists? The following formula provides the appropriate answer:

$$r = 2\left(\frac{\sigma}{\delta}\right)^2 \left[t_{\alpha(df)} + t_{\beta(df)} \right]^2$$

where r is the number of replicates per treatment, σ is the true standard deviation of the experimental units, df are the degrees of freedom for the experimental error in the proposed ANOVA, α is the desired level of significance, β is the desired probability of a Type II error (β = 1-power of the test), $t_{\alpha(df)}$ and $t_{\beta(df)}$ are two- and one-tailed tabular t values, respectively, with df degrees of freedom at the α and β probabilities, respectively.

An example will best illustrate the application of this procedure. Consider the Latin square experiment of the previous section as a basis for the example. Suppose that the researcher wishes to conduct future studies (4 treatments) capable of detecting differences between treatment means of 50 g in dry matter weight, at the 5% level of significance and with a 95% power of the test. The previous experiment will be used to provide an estimate of experimental error ($\hat{\sigma} = (MS_E)^{1/2} = 22$). Since r must be known to find the correct t values, the solution to the equation is iterative. That is, we initially make a guess and solve the equation. Then using the solution to aid with further guesses we repeat the process until our guess equals the solution. As an initial guess, we try r = 4, or df_E = 6.

$$r \cong 2\left(\frac{22}{50}\right)^2 (2.447 + 1.943)^2$$

$$\cong 7.5$$

The solution would indicate that our guess was too small. The 7.5 would suggest that r = 8.

As a second guess try r = 8, and df_E equal 18.

$$r \cong 2\left(\frac{22}{50}\right)^2 (2.101 + 1.734)^2$$

$$\cong 5.69$$

At this point further calculation is unnecessary, since we must have 2 complete squares to achieve the minimum standard the researcher suggested. If the researcher now conducts his experiment (assuming his estimate of σ is correct), he is assured that should the real treatment differences be 50 g or greater, he is at least 95% sure of detecting the difference as significant (P<0.05).

PROBLEMS

8.1. A 9 member taste panel was asked to score hamburgers for tenderness on a scale from 1 to 9. The study was concerned with methods of preparing the hamburgers which were believed to influence tenderness. The numeric scale from 1 to 9 was used where 1 equaled "extremely tender" and 9 equaled "extremely tough."

			Method			
Panelist	A	B	C	D	E	F
1	3	4	3	5	3	2
2	1	2	1	4	2	1
3	1	1	1	3	2	2
4	3	2	2	4	4	4
5	7	6	5	9	5	6
6	5	6	8	7	9	5
7	6	5	3	5	4	5
8	3	1	2	6	5	1
9	5	3	3	5	5	2

(a) Analyze the data as an RCB design with panelists representing complete replicates. Test the hypothesis of no difference between methods of preparation.

(b) Suppose that the score for panelist 8, method D was lost during the experiment. Use the missing data techniques to estimate the missing observation and complete the ANOVA.

8.2. A digestion trial was conducted to evaluate dry matter digestibility of 4 rations for cattle. Four animals were used for the study with each animal assigned to each ration in a different time period in a Latin square design. The letters in the table correspond to the ration and the number is the dry matter digestion coefficient.

			Animal		
		1	2	3	4
	I	76 (B)	66 (D)	67 (A)	64 (C)
	II	64 (D)	70 (C)	71 (B)	65 (A)
Period	III	72 (A)	73 (B)	66 (C)	63 (D)
	IV	68 (C)	70 (A)	62 (D)	66 (B)

(a) Complete the ANOVA, testing the hypothesis of no difference among rations.

(b) Use the Duncan's multiple range test to determine which ration means are significantly different.

8.3. Suppose that the researchers in problem 8.1 wish to compute the required sample size for future studies. Use the information from the ANOVA and the following specifications from the researchers to determine the number of panelists needed in future studies. The researchers wish to be able to detect a difference of 1 unit on their tenderness scale, with 95% certainty, at the 5% level of significance.

REFERENCES

Additional examples of application of RCB and Latin squares may be found in:

OSTLE, B. and MENSING, R.W. 1975. Statistics in Research. Iowa State Univ. Press, Ames.

STEEL, R.G.D. and TORRIE, J.H. 1960. Principles and Procedures of Statistics. McGraw-Hill Book Company, New York.

For a discussion of design aspects and variations in the basic designs see:

COCHRAN, W.G. and COX, G.M. 1957. Experimental Designs, 2nd Edition. John Wiley & Sons, New York.

LI, C.C. 1964. Introduction to Experimental Statistics. McGraw-Hill Book Company, New York.

YATES, F. 1933. The analysis of replicated experiments when the field results are incomplete. Empire J. Exp. Agric. *1*, 129–142.

Examples of RCB and Latin square applications in recent literature

ARNDT, D.L., RICHARDSON, C.R., ALBIN, R.C. and SHERROD, B. 1980. Digestibility of chemically treated cotton plant byproduct and effect on mineral balance, urine volume and pH. J. Anim. Sci. *51*, 215.

SEMAN, D.L., OLSON, D.G. and MANDIGO, R.W. 1980. Effect of reduction and partial replacement of sodium on bologna characteristics and acceptability. J. Food Sci. *45*, 1116.

STANHOPE, D.L., HINMAN, D.D., EVERSEN, D.O. and BULL, R.C. 1980. Digestibility of potato processing residue in beef cattle finishing diets. J. Anim. Sci. *51*, 202.

STONE, M.B. and CAMPBELL, A.M. 1980. Emulsification in systems containing soy protein isolates, salt and starch. J. Food Sci. *45*, 1713.

9

Factorial Experiments

As an introduction to this chapter, two questions need answering. First, what are factorial experiments? Second, why use factorial experiments?

In experimental design a particular treatment (independent variable) is referred to as a *factor*. In the examples presented in previous chapters we have been concerned only with single factor experiments. A factorial experiment is a set of treatments made up of 2 or more factors. Each factor is present in the experiment at 2 or more levels (generally quantitative), such that all possible combinations of the levels of the 2 factors are present. For example, an agronomist might wish to examine the effects of different amounts and sources of nitrogen on crop yield. The first factor would be amount, say 50, 100, and 150 lb per acre (56.2, 112.5, and 168.8 kg/ha) of actual nitrogen, while the second factor is the source, for example, urea or ammonia nitrogen, thus resulting in a 3 × 2 factorial experiment.

The simplest factorial is the 2 × 2 factorial treatment arrangement. That is, 2 factors each at 2 levels, making 4 treatments in total. Each number represents a factor and the digit 2 indicates the number of levels of the factor.

The major portion of this chapter will be devoted to the 2 × 2 factorial to serve as a basis for understanding the details of the factorial analysis and interpretation. The factorial experiment is not limited (in theory) with respect to either the number of factors or the number of levels for a factor. However, in practical application, a 4 factor experiment, for example with the levels 2 × 4 × 5 × 3, would result in 120 treatment combinations, which in many experimental situations would be too cumbersome for a single experiment.

Note that we have not referred to the factorial as a design. In fact, the factorial experiment is nothing other than a particular set of treatment combinations. Any of the 3 experimental designs presented earlier (CRD, RCB, or Latin square) may contain a factorial treatment combination. You should recognize, however, that in the case of the Latin square the dimensions of the factorial arrangement would have to be kept small or the square would become unwieldy. For example, even a modest factorial, such as a 2 × 2 × 3 = 12, would require 144 experimental units (12 × 12 Latin square)

since the number of rows and columns must equal the number of treat
ments. It is true, however, that the factorial treatment arrangement ha:
resulted in the development of certain experimental designs which are use(
almost exclusively with factorial treatment arrangements.

Now that we have considered what a factorial experiment is, let's turn ou
attention to why we conduct factorial experiments. There are essentiall:
two reasons why factorial experiments are conducted. They are simpl;
"realism" and "efficiency." Let us examine each aspect individually.

For years the single factor experiment was the standard research metho(
(and it still is in many areas of research). It is an experiment conducted witl
all factors held constant except the one to be studied. The analysis an(
interpretation of such experiments were generally simple and straightfor
ward. However, since researchers and users of research information like t(
generalize the results of such experiments, they were frequently mislead
ing. The reason was that, in practice or in additional research, when factor
being held constant were fixed at different levels, the treatments responde(
differently. So what at first appeared to be a sound conclusion turned out t(
be misleading or clearly wrong when other conditions were changed. Thi
phenomenon of the response of one factor depending on the level of th(
second factor is known as an "interaction."

Since in biology one should at least consider the possibility of interactioı
among factors, the single factor study is generally inadequate. The factoria
experiment, by including independent combinations of the levels of al
factors, permits the researcher to test hypotheses concerning interactioı
effects. Therefore, it should be obvious to most biologists that the results o
the analysis of factorial treatment combinations would be more realisti
biologically than the results of single factor studies.

The ability to test for the presence of interaction effects is sufficien
justification in biology for conducting factorial experiments. However, iı
addition, factorials, in general, more efficiently use available experimenta
material. Every experimental unit provides some information about eacl
factor included in the experiment as well as information about all th
interactions. Suppose, for example, we are conducting a $2 \times 2 \times 2$ factoria
experiment. If we simply call the factors A, B, and C, then in the factoria
analysis we will test hypotheses about each of the factors (main effects
plus 3 2-factor interactions (AB, AC, and BC) and the 3-factor interaction

In biology, the testing and interpretation of interactions is likely to be o
the greatest importance. It should be noted that in the absence of interac
tions, a single $2 \times 2 \times 2$ factorial experiment using n experimental unit
will provide as much information about the main effects as 3 separate singl
factor experiments each using n units. In addition, the researcher could tes
the hypothesis of no-interaction rather than assuming no-interaction.

FACTORIAL TREATMENT ARRANGEMENTS

Before we consider the factorial analysis in detail we need to study th
notation and terminology used for factorial experiments. Although th

notation presented is not universal, it is probably the most common. Factors are denoted by single capital letters, while levels are denoted by lower case letters with subscripts to identify the level. For example, in a 3×2 factorial, we have factor A with 3 levels (a_1, a_2, a_3) and factor B with 2 levels (b_1, b_2). Therefore the 6 treatment combinations would be represented as a_1b_1, a_1b_2, a_2b_1, a_2b_2, a_3b_1, a_3b_2. The treatment a_2b_1 refers to the combination resulting from the application of the second level of factor A with the first level of factor B.

We have already used the terms "main effect" and "interaction" of factors. These two terms are commonly associated with discussions of factorial experiments. A clear understanding of their meaning is essential to the proper use and interpretation of factorial experiments. Let's first examine the concept of factor interactions and then we will turn our attention to the main effects.

Two factors A and B are said to "interact" when the responses to the levels of factor A are not the same at all levels of factor B. This is also the same as saying the responses to the levels of factor B are not the same at all levels of factor A. What we are saying is that factors A and B are not independent. The effects of factor A depend on the level of factor B (same as saying the effects of factor B depend on the level of factor A).

The minimum requirement for the evaluation of an interaction is the 2×2 factorial experiment. It is also this simplest case which most readily lends itself to explaining what is meant by interaction of factors A and B. Symbolically the treatments would be presented as follows:

	A	
B	a_1	a_2
b_1	a_1b_1	a_2b_1
b_2	a_1b_2	a_2b_2

Now suppose that the response to treatment a_1b_1 is some value, Y. Further suppose that treatment a_2b_1 results in the response Y plus some constant c and that the a_1b_2 treatment response is Y plus some constant k. The measured response would have the values:

Measured Response

	A	
B	a_1	a_2
b_1	Y	Y + c
b_2	Y + k	?

The question of interest is, then, what is the response for treatment a_2b_2? If, for example, we find the response is $Y + c + k$, then the factors are said to be independent (no-interaction). In statistics, we would commonly say that the effects of factors A and B are "additive." In this case one should note that when going from the a_1 to a_2 level, the effect is $+c$ regardless of the level of B and when we go from the b_1 to b_2 level, the effect is $+k$ regardless of the level of A. Thus the response to the a_2b_2 treatment combination is equal to the response to the a_1b_1 treatment plus the sum of the 2 constants $(c + k)$.

Now suppose that $a_2b_2 >$ or $< Y + c + k$. In this example we would say that factors A and B "interact" (their responses are nonadditive). Note that there are two distinct kinds of interaction. If the response for a_2b_2 were greater than $Y + c + k$, then somehow the combination at the higher level of each factor resulted in a greater response than was indicated by the simple effect observed at the lower levels. If the response for a_2b_2 were less than $Y + c + k$, then the combination at the higher levels resulted in a lesser response than was indicated by the simple effects at the lower levels.

In biology the first case is not unlike the phenomenon known as "synergism," while the second case might be referred to as an "inhibitory response." The student of statistics should be cautioned that the presence of a "statistical interaction" is simply an evaluation of the relation among a particular combination of 4 treatment means. In fact, the possibility of a biological interaction need not exist at all in a factorial experiment.

Some numerical examples and figures should help to clarify the concepts of interaction and independence. In Table 9.1 and Fig. 9.1 three examples are presented. Example (a) represents the case of no-interaction (independence or additivity). Note that the case of no-interaction results in parallel response lines, while interaction is indicated by nonparallel lines. Graphic presentations are frequently the simplest way to present the results of a factorial experiment when interactions are present. These figures also help to illustrate another way in which factorials are sometimes used. That is to test for parallelism of two or more responses observed over several treatment levels or time periods.

TABLE 9.1.   EXAMPLES OF NO-INTERACTION (a) AND INTERACTIONS (b AND c) IN 2 × 2 TABLES

B	A a_1	a_2	A a_1	a_2	A a_1	a_2
b_1	2	5	2	5	2	5
b_2	3	6	3	8	3	4
	(a)		(b)		(c)	

Now let us turn our attention to the main effects. The effect of the "i"th level of the one factor averaged over all the levels of the second factor is called a "main effect." The individual effects at each level are referred to as "simple effects." Table 9.2 presents the simple and main effects of the data in Table 9.1.

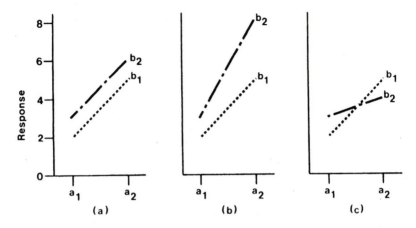

FIG. 9.1.   FIGURES REPRESENT NO-INTERACTION (a) AND INTERACTION EFFECT (b AND c) FOR 2 × 2 FACTORIALS

TABLE 9.2.   ILLUSTRATION OF SIMPLE AND MAIN EFFECTS

(a)	A		Simple Effects of A
B	a_1	a_2	$a_2 - a_1$
b_1	2	5	+3
b_2	3	6	+3
Simple Effects of B			+3 Main Effect of A
$b_2 - b_1$	+1	+1	+1 Main Effect of B
(b)	A		Simple Effects of A
B	a_1	a_2	$a_2 - a_1$
b_1	2	5	+3
b_2	3	8	+5
Simple Effects of B			+4 Main Effect of A
$b_2 - b_1$	+1	+3	+2 Main Effect of B
(c)	A		Simple Effects of A
B	a_1	a_2	$a_2 - a_1$
b_1	2	5	+3
b_2	3	4	+1
Simple Effects of B			+2 Main Effect of A
$b_2 - b_1$	+1	-1	+0 Main Effect of B

Note that in Table 9.2 (a) where the data show no-interaction, the simple effects are the same within a factor. The differences between simple effects within a factor are a measure of the interaction of the two factors. Since the interaction of A with B is the same as saying the interaction of B with A we note that differences between simple effects for each table are the same for both factors A and B; that is, the difference between simple effects for each factor is 0 in (a), 2 units in (b), and 2 units in (c).

The presence (or absence) of main effects is independent of the presence (or absence) of interactions, that is, we may observe interaction with or without main effects and vice versa. Likewise we may observe one main effect and not the other. For the three examples in Table 9.2 we have in (a) no-interaction but main effects of A and B, (b) interaction and main effects of A and B, and (c) interaction and main effect of A, but not B. Several other combinations are possible, for example, no main effects but interaction.

The student should recognize that in the examples given here we did not judge the significance or nonsignificance of the effects. The purpose of the factorial ANOVA is to determine which effects are significant. In actual practice the magnitude of the effects given as examples may or may not be large enough to be judged significant. In an experiment, we would expect to find variation such that the observed effects would seldom equal zero. However, when the observed effects are small enough to be attributed to experimental variation, we accept the null hypothesis of no-effect.

One additional way of looking at factorials is frequently useful to students. Since factorials are in fact simply a particular arrangement of treatments, we might attempt to look at main effects and interactions as orthogonal linear contrast just as we did with the arrangements of treatments in Chapter 7.

For a 2×2 factorial there would be 4 treatments and 3 degrees of freedom. Therefore, 3 orthogonal linear contrasts can be proposed to partition the treatment sums of squares into the A main effect, B main effect, and the AB interaction effect. The constants for partitioning the 2×2 factorial are presented in Table 9.3. Although more general computing formulas will be given later in the chapter, remember that the SS $= \dfrac{n(\Sigma c_i \bar{Y}_i)^2}{\Sigma c_i^2}$.

TABLE 9.3.   CONSTANTS (c_i) FOR 2×2 ORTHOGONAL LINEAR CONTRAST

Effect	Treatment			
	a_1b_1	a_1b_2	a_2b_1	a_2b_2
A	$-\frac{1}{2}$	$-\frac{1}{2}$	$+\frac{1}{2}$	$+\frac{1}{2}$
B	$-\frac{1}{2}$	$+\frac{1}{2}$	$-\frac{1}{2}$	$+\frac{1}{2}$
AB	$+\frac{1}{2}$	$-\frac{1}{2}$	$-\frac{1}{2}$	$+\frac{1}{2}$

It should be of interest to compute the value of the contrast for each of the examples given in Table 9.1. For Table 9.1 (a) we compute (the bar over the treatment combination representing the mean):

A Main Effect:

$$C_A = -\tfrac{1}{2}\overline{(a_1b_1)} - \tfrac{1}{2}\overline{(a_1b_2)} + \tfrac{1}{2}\overline{(a_2b_1)} + \tfrac{1}{2}\overline{(a_2b_2)}$$

$$= -\tfrac{1}{2}(2) - \tfrac{1}{2}(3) + \tfrac{1}{2}(5) + \tfrac{1}{2}(6)$$

$$= +3$$

B Main Effect:

$$C_B \quad = \quad -\tfrac{1}{2}(2) \; + \; \tfrac{1}{2}(3) \; - \; \tfrac{1}{2}(5) \; + \; \tfrac{1}{2}(6)$$

$$= \; +1$$

AB Interaction:

$$C_{AB} \quad = \quad +\tfrac{1}{2}(2) \; - \; \tfrac{1}{2}(3) \; - \; \tfrac{1}{2}(5) \; + \; \tfrac{1}{2}(6)$$

$$= \; 0$$

For Table 9.1(b):

$$C_A \quad = \quad -\tfrac{1}{2}(2) \; - \; \tfrac{1}{2}(3) \; + \; \tfrac{1}{2}(5) \; + \; \tfrac{1}{2}(6)$$

$$= \; +4$$

$$C_B \quad = \; +2$$

$$C_{AB} \quad = \; +1$$

For Table 9.1(c):

$$C_A \quad = \; +2$$

$$C_B \quad = \; 0$$

$$C_{AB} \quad = \; -1$$

Note that the interaction effect is $\tfrac{1}{2}$ of the difference between the simple effects given in Table 9.2 [i.e., (b) $C_{AB} = \tfrac{1}{2}(3 - 1)$ or $\tfrac{1}{2}(2)$]. Also note that if the constants are properly constructed, the sign of the effect has a meaning. For example, here the positive main effect indicates the second level was greater than the first, while a negative sign indicates the second level was less than the first. A positive interaction indicates a response of a_2b_2 that is greater than additivity, while the negative sign indicates that the a_2b_2 response was less than additive.

THE LINEAR ADDITIVE MODEL

At this point we should turn our attention to statistical models associated with factorial analysis. Since the factorial treatment arrangement may be used in any of the 3 designs (CRD, RCB, and Latin square) presented in the last two chapters, the complete model will vary depending on the experimental design. If we first consider the factorial model associated with the completely random design, then the necessary changes should be obvious for the other designs.

The linear additive model for the completely random design with t treatments and n replication for each treatment is:

$$Y_{ij} = \mu + \tau_i + \epsilon_{ij}$$

where i = $1, \cdots, t$
        j = $1, \cdots, n$
        μ is the true mean
        τ_i is the true effect of the "i"th treatment
and ϵ_{ij} is the true effect of the jth experimental unit on the "i"th treatment assumed to be NID $(0, \sigma^2)$.

Now suppose that we know that the t treatments are arranged factorially. That is, all combinations of factors A and B (t equals ab) are present in the study. The treatment effect τ_i in the preceding model is now known to equal $\alpha_i + \beta_j + (\alpha\beta)_{ij}$, where α and β represent the true main effects of factors A and B, while $(\alpha\beta)$ represents the AB interaction effect. If we now write the complete model for the CRD with factorial arrangement of treatments, we have:

$$Y_{ijk} = \mu + \alpha_i + \beta_j + (\alpha\beta)_{ij} + \epsilon_{ijk}$$

where i = $1, \cdots, a$
        j = $1, \cdots, b$
        k = $1, \cdots, n$
        μ is the true mean
        α_i is the true effect of the jth level of factor A
        β_j is the true effect of the jth level of factor B
        $\alpha\beta_{ij}$ is the true effect of the interaction of the "i"th level of factor A with the jth level of factor B
and ϵ_{ijk} is the true effect of the kth experimental unit on the ijth treatment, assumed to be NID $(0, \sigma^2)$.

When the treatment combinations are arranged factorially for an RCB or Latin square design, we substitute the factorial effects for the treatment effects in the appropriate model. For example, in an RCB design if τ_i represents treatment and γ_j represents block effects, we would write the model as:

$$Y_{ij} = \mu + \tau_i + \gamma_j + \epsilon_{ij}$$

If the treatment effects were arranged factorially, we would write:

$$Y_{ijk} = \mu + \alpha_i + \beta_j + (\alpha\beta)_{ij} + \gamma_k + \epsilon_{ijk}$$

For the Latin square design, if γ and δ represented row and column effects, respectively, for the factorial arrangement we would write:

$$Y_{ijkl} = \mu + \alpha_i + \beta_j + (\alpha\beta)_{ij} + \gamma_k + \delta_j + \epsilon_{ijkl}$$

As before, a complete description of the model will require identification of the effects as fixed or random. Since consequences of an effect being fixed or random influence how the expected mean squares are written and how F tests are computed, we will discuss fixed, random, and mixed models at that time.

THE 2 × 2 FACTORIAL EXPERIMENT

In this section we will examine in detail the analysis of a 2 × 2 factorial experiment. We will consider the computation of the sums of squares, the expected mean squares, appropriate test of significance, and finally the interpretation and presentation of the results.

As an example of a 2 × 2 factorial let us consider an experiment designed to study appetite response of rats fed low or high levels of protein (factor A) and low or high levels of fat (factor B). The measured variable is food intake expressed as the percentage above the maintenance requirement. The design of the experiment was completely random with 7 rats of approximately uniform weight assigned to each diet. The data are presented in Table 9.4. In Table 9.5 we have presented the treatment totals arranged in a 2 × 2 table with marginal totals corresponding to the main effects and the grand total in the lower right corner. The totals presented in Table 9.5 are the ones necessary for the computation of the sums of squares for the 2 × 2 factorial.

TABLE 9.4. FOOD INTAKE AS THE PERCENTAGE GREATER THAN MAINTENANCE REQUIREMENT

Protein (i = 1,. . ., a)	Low		High	
Fat (j = 1,. . ., b)	Low	High	Low	High
	31	48	46	24
Replication (k = 1,...,n)	42	51	51	27
	59	59	32	27
	57	49	50	29
	55	42	34	47
	43	50	52	35
	35	52	46	48
$Y_{ij\cdot}$	322	351	311	237
$\bar{Y}_{ij\cdot}$	46.0	50.1	44.4	33.9

TABLE 9.5. 2-WAY TABLE OF TOTALS FOR COMPUTATION OF THE SUMS OF SQUARES

Fat	Protein		$Y_{\cdot j\cdot}$
	Low	High	
Low	322	311	633
High	351	237	588
$Y_{i\cdot\cdot}$	673	548	1221

TABLE 9.6. DEFINITIONAL AND COMPUTATIONAL FORMULAS FOR THE 2 × 2 FACTORIAL EXPERIMENT

Source of Variation	Degrees of Freedom	Sums of Squares Definitional	Sums of Squares Computational
Treatment	$ab - 1$	$n\Sigma(\bar{Y}_{ij\cdot} - \bar{Y}...)^2$	$\dfrac{\Sigma Y_{ij}^2}{n} - \dfrac{Y^2...}{abn}$
A	$a - 1$	$bn\Sigma(\bar{Y}_{i\cdot\cdot} - \bar{Y}...)^2$	$\dfrac{\Sigma Y_{i\cdot\cdot}^2}{bn} - \dfrac{Y^2...}{abn}$
B	$b - 1$	$an\Sigma(\bar{Y}_{\cdot j\cdot} - \bar{Y}...)^2$	$\dfrac{\Sigma Y_{\cdot j}^2}{an} - \dfrac{Y^2...}{abn}$
AB	$(a - 1)(b - 1)$	$n\Sigma(\bar{Y}_{ij\cdot} - \bar{Y}_{i\cdot\cdot} - \bar{Y}_{\cdot j\cdot} + \bar{Y}...)^2$	$SS_T - SS_A - SS_B$
Within	$ab(n - 1)$	$\Sigma(\bar{Y}_{ijk} - \bar{Y}_{ij\cdot})^2$	$SS_{Total} - SS_T$
Total	$nab - 1$	$\Sigma(Y_{ijk} - \bar{Y}...)^2$	$\Sigma Y_{ijk}^2 - \dfrac{Y^2...}{abn}$

As in the previous two chapters let us examine the definitional and computational formulas for the sums of squares (Table 9.6). First examine the table and then we will proceed to compute the SS by the computational formulas as follows:

$$SS_{Total} = (31^2 + 42^2 + \ldots + 48^2) - \frac{1221^2}{(2)(2)(7)}$$

$$= 2894.68$$

$$SS_{Treatment} = \frac{322^2 + 351^2 + 311^2 + 237^2}{7} - \frac{1221^2}{(2)(2)(7)}$$

$$= 1009.25$$

$$SS_A = \frac{673^2 + 548^2}{(2)(7)} - \frac{1221^2}{(2)(2)(7)}$$

$$= 558.04$$

$$SS_B = \frac{633^2 + 588^2}{(2)(7)} - \frac{1221^2}{(2)(2)(7)}$$

$$= 72.34$$

$$SS_{AB} = 1009.25 - 558.04 - 72.34$$

$$= 378.87$$

$$SS_{Within} = 2894.68 - 1009.25$$

$$= 1885.43$$

Before we write out the complete ANOVA, let's consider the expected mean squares for the 2-factor factorial (Table 9.7). It is necessary that the reader understand how to use the E(MS) to construct appropriate F ratios. In this case there are 3 null hypotheses to be tested; thus there are 3 F ratios to be computed.

TABLE 9.7. EXPECTED MEAN SQUARES 2-FACTOR FACTORIAL COMPLETELY RANDOM DESIGN

Source	Expected Mean Squares		
	Fixed	Random	Mixed A Fixed, B Random
A	$\sigma^2 + \dfrac{nb}{a-1}\Sigma\alpha_i^2$	$\sigma^2 + n\sigma_{\alpha\beta}^2 + nb\sigma_\alpha^2$	$\sigma^2 + n\sigma_{\alpha\beta}^2 + \dfrac{nb}{a-1}\Sigma\alpha_i^2$
B	$\sigma^2 + \dfrac{na}{b-1}\Sigma\beta_j^2$	$\sigma^2 + n\sigma_{\alpha\beta}^2 + na\sigma_\beta^2$	$\sigma^2 + na\sigma_\beta^2$
AB	$\sigma^2 + \dfrac{n}{(a-1)(b-1)}\Sigma(\alpha\beta)_{ij}^2$	$\sigma^2 + n\sigma_{\alpha\beta}^2$	$\sigma^2 + n\sigma_{\alpha\beta}^2$
Within	σ^2	σ^2	σ^2

To illustrate the problem, consider the random model. The first hypothesis to be tested concerns the AB interaction, written as H_o: $\sigma_{\alpha\beta}^2 = 0$. The student should recall that the F ratio should equal 1 if the null hypothesis is true, that is, the numerator and the denominator of the F ratio should be identical except for that portion of the numerator mean square associated with the term being tested by the null hypothesis. Since the E(MS) for the AB interaction is $\sigma^2 + n\sigma_{\alpha\beta}^2$, the denominator mean square should have an expected value of σ^2. Thus the F ratio to test the null hypothesis H_o: $\sigma_{\alpha\beta}^2 = 0$ is

$$F = \frac{\sigma^2 + n\sigma_{\alpha\beta}^2}{\sigma^2} = \frac{MS_{AB}}{MS_{Within}}$$

Now consider the main effects for the random model. The two null hypotheses are H_o: $\sigma_\beta^2 = 0$ and H_o: $\sigma_\alpha^2 = 0$. Note that in each case the expected value for the main effect mean square includes both the experimental variance (σ^2) and the interaction variance ($\sigma_{\alpha\beta}^2$) in addition to the main effect component of variance. The appropriate F ratios for the two main effects in the random model would be:

$$F = \frac{\sigma^2 + n\sigma_{\alpha\beta}^2 + na\sigma_\beta^2}{\sigma^2 + n\sigma_{\alpha\beta}^2} = \frac{MS_B}{MS_{AB}}$$

$$\text{and } F = \frac{\sigma^2 + n\sigma^2_{\alpha\beta} + nb\sigma^2_{\alpha}}{\sigma^2 + n\sigma^2_{\alpha\beta}} = \frac{MS_A}{MS_{AB}}$$

Note that only for the fixed model is the within MS the appropriate denominator for all 3 null hypotheses. Table 9.8 lists the appropriate F ratios for all three 2-way models. For higher order factorial experiments, cases exist for random and mixed models where no appropriate single denominator mean square exists for the desired F ratio. In these cases, approximate F ratios may be constructed by adding several MS to create the appropriate combination for the test of hypothesis.

TABLE 9.8.   F RATIOS FOR THE 2-WAY FACTORIAL

	F Ratios		
Sources	Fixed A and B Fixed	Random A and B Random	Mixed A Fixed, B Random
A	MS_A/MS_W	MS_A/MS_{AB}	MS_A/MS_{AB}
B	MS_B/MS_W	MS_B/MS_{AB}	MS_B/MS_W
AB	MS_{AB}/MS_W	MS_{AB}/MS_W	MS_{AB}/MS_W

The E(MS) of Table 9.7 are very closely associated with the E(MS) for the RCB analysis given in the previous chapter. While we are studying E(MS) it would probably be helpful to many readers to examine the similarity between the two sets of expected mean squares. It is very easy to go from the E(MS) of Table 9.7 to those in Table 8.4. We would first need to drop out the line associated with the within source of variation in Table 9.7. This is not to say that experimental error does not exist for the RCB, but only that we cannot measure it directly since only one observation exists for each block and treatment combination. Now relabel the A, B, and AB effects as treatment, blocks, and error, respectively. At this point, we now have the expected mean squares for the RCB design except for the assumption of no interaction. Note that for the fixed model, if block by treatment interaction is present, the expected F ratio is

$$\frac{\sigma^2 + bn\dfrac{\Sigma\alpha^2}{a-1}}{\sigma^2 + n\dfrac{\Sigma(\alpha\beta)^2}{(a-1)(b-1)}}$$

Clearly this is an inappropriate test of the hypothesis that $\Sigma\alpha^2 = 0$. Even though the F test for treatment effects is appropriate for the random and fixed models, the interpretation of the main effects would be difficult. For these reasons we use the RCB design only when it is reasonable to assume no interaction of blocks and treatments. To obtain the expected mean squares in Table 8.4, the only remaining step is to delete the term associated with the interaction components.

Now let's return to our example and complete the ANOVA. For our example both factor A and B are fixed. That is, the researcher specifically selected a high and low level of both protein and fat. Therefore the hypotheses to be tested are H_o: $\Sigma(\alpha\beta)^2 = 0$, H_o: $\Sigma\beta^2 = 0$, and H_o: $\Sigma\alpha^2 = 0$. As shown in Table 9.8, the MS_{Within} is the appropriate denominator for all the F ratios. The complete ANOVA is given in Table 9.9. Note that the treatment and total sources of variation are required for the computational procedures, but could be omitted from the ANOVA table. The A, B, and AB sources are indented to show that they represent a partitioning of the treatment SS.

TABLE 9.9.   EXAMPLE OF A 2 × 2 FACTORIAL ANOVA FOR THE DATA OF TABLE 9.4

Source	df	SS	MS	F
Treatments	3	1009.25	—	—
A	1	558.04	558.04	7.10
B	1	72.34	72.34	<1
AB	1	378.87	378.87	4.82
Within	24	1885.43	78.56	—
Total	27	2894.68	—	—

The 5% critical F value from Appendix Table D.1 with 1 and 24 df is 4.26. Therefore the F statistics for AB interaction and A main effects indicate rejection of the null hypothesis. The F statistic for the B main effect is simply marked < 1 since an F ratio less than 1 would not lead to rejection of the null hypothesis. The AB interaction means that the difference between low and high protein is not the same for the two fat levels (the same as saying that the difference between the two fat levels was not the same at different protein levels). It is this interaction effect which must be examined before anything can be said about the main effect of A.

The results of a significant interaction are generally presented either as a 2-way table of means or in graph form. Although both approaches present the same data, one approach may be more appealing to some readers or in some situations than the other. Let us look at both approaches. Table 9.10 presents the means and Fig. 9.2 is a graph of the interaction. In the table of means the standard error of the mean is also reported. Although individual SEM could have been computed, those in the table have been computed from the MS_{Within}

$$SEM = \left(\frac{MS_W}{n} \right)^{1.2} = \left(\frac{78.56}{7} \right)^{1.2} = 3.4$$

TABLE 9.10.   2-WAY TABLE OF TREATMENT MEANS FOR EXAMINATION OF SIGNIFICANT INTERACTION EFFECT

Mean (± SEM) Food Intake (%) Above Maintenance Requirement		
	Protein	
Fat	Low	High
Low	46. ± 3.4	44. ± 3.4
High	50. ± 3.4	34. ± 3.4

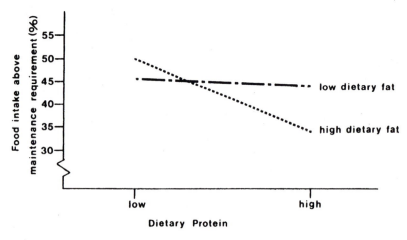

FIG. 9.2.   TREATMENT MEANS FOR SIGNIFICANT PROTEIN BY FAT INTERACTION

The means have been reported to the units position, followed by a decimal to indicate that the units position represents a significant figure. This decision was not based on the original data, but on the magnitude of the standard error. The 95% confidence limits are approximately ± 2 SEM; therefore, the low-low treatment limits are ~ 39.2 to 52.8. A good "rule of thumb" is that the last significant digit of the mean (or any statistic) corresponds to the first non-zero digit of the SEM (or standard error of the statistic).

The table of means lends itself to the examination of simple effects or to the additivity of the effects. The drop in intake from low to high protein was 2 points (percentage points) at the low fat level, while at the high fat level the drop was 16 points. It is this difference of 14 percentage points between simple effects that was found significant.

We can also use the table of means to look at additivity of the simple effects. If we use the high-high treatment as the starting treatment, the shift to low fat resulted in a 10 point increase in intake, while the shift to low protein resulted in a 16 point increase. If the effects were additive, we would expect a 26 point (10 + 16) increase in intake from the high-high to the low-low treatment for a treatment mean of 60. The difference between the observed mean of 46 and the expected mean of 60 is the (14 percentage points) significant interaction effect.

The interaction can also be seen by examination of Fig. 9.2. If no interaction were present, then the two lines would not differ significantly from parallel. Since in this case the interaction was found significant, the nature of the nonparallelism is helpful in understanding the interaction. We can see from Fig. 9.2 that the effect of going from a low to a high protein diet depends on the level of fat in the diet. On the high fat diets, the food intake was greatly reduced on the high protein diet as compared to the low protein diet, while on the low fat diets, intake was relatively unaffected by the protein level in the diet.

OTHER FACTORIAL EXPERIMENTS

To this point in Chapter 9 we have considered only the simplest factorial experiment (2×2). Factorial experiments with more than 2 levels (i.e., $2 \times 3, 3 \times 4$, etc.) and experiments with more than 2 factors (i.e., $2 \times 2 \times 2, 2 \times 3 \times 4$, etc.) are general extensions of the simple 2×2 factorial. In other words, the concept of testing main effects and interaction effects is the same. Interpretations may be more difficult, but only because there are more main effects and interactions (perhaps of higher order) requiring interpretation. However, in biology such models may be more realistic than models which fail to include these interactions.

Now consider an experiment with 3 treatment factors. A greenhouse experiment was conducted to study the effects of potassium fertilization and soil moisture on 3 varieties of soybeans. Among the dependent variables measured was seed size, measured as weight per 100 seeds. The experiment was conducted in 3 randomized complete blocks, controlling for location effects in the greenhouse.

Table 9.11 presents the 24 treatment totals for 3 replications. In addition, 2-way tables of totals and margin totals have been presented (Table 9.12). These tables contain the necessary totals for computing the various sums of squares for the 3-factor factorial. Individual observations have not been presented. However, the total, error, and block sums of squares were computed in the usual manner from the data not present here.

The linear additive model for the experiment could be presented as follows:

$$Y_{ijkl} = \mu + \delta_l + \alpha_i + \beta_j + \gamma_k + (\alpha\beta)_{ij} + (\alpha\gamma)_{ik} + (\beta\gamma)_{jk} + (\alpha\beta\gamma)_{ijk} + \epsilon_{ijkl}$$

where i $= 1, \cdots m$
     j $= 1, \cdots v$
     k $= 1, \cdots k$
     l $= 1, \cdots r$

TABLE 9.11.   TREATMENT TOTALS ($r = 3$) FOR THE SEED SIZE TREATMENT

Moisture (M)	Variety (V)	Potassium (K)			
		0	1	2	3
Deficient	Blackhawk	36.9	37.7	37.3	38.7
	Lincoln	33.6	34.7	36.7	36.3
	Hawkeye	42.3	41.6	43.9	41.1
Adequate	Blackhawk	39.6	38.7	43.8	51.9
	Lincoln	42.9	45.3	47.4	48.6
	Hawkeye	46.5	48.9	54.8	58.6

TABLE 9.12.   2-WAY TABLES OF TOTALS FOR THE SEED SIZE EXPERIMENT

		Variety (v = 3)			
		v_1	v_2	v_3	Sum
Moisture	m_1	150.6	141.3	168.9	460.8
(m = 2)	m_2	174.0	184.2	208.8	567.0
Sum		324.6	325.5	377.7	1027.8

		Potassium (k = 4)				
		k_0	k_1	k_2	k_3	Sum
Moisture	m_1	112.8	114.0	117.9	116.1	460.8
(m = 2)	m_2	129.0	132.9	146.0	159.1	567.0
Sum		241.8	246.9	263.9	275.2	1027.8

		Potassium (k = 4)				
		k_0	k_1	k_2	k_3	Sum
Variety	v_1	76.5	76.4	81.1	90.6	324.6
(v = 3)	v_2	76.5	80.0	84.1	84.9	325.5
	v_3	88.8	90.5	98.7	99.7	377.7
Sum		241.8	246.9	263.9	275.2	1027.8

α_i is the effect of the "i"th moisture level

β_j is the effect of the jth variety

γ_k is the effect of the kth potassium

$(\alpha\beta)_{ij}$, $(\alpha\gamma)_{ik}$, $(\beta\gamma)_{jk}$ and $(\alpha\beta\gamma)_{ijk}$ represent the first and second order interactions effects for the α, β, and γ factors

and ϵ_{ijkl} represents the random error effects, assumed to be NID $(0,\sigma^2)$.

The computation of the sums of squares for this RCB design with the 24 treatments arranged in a $2 \times 3 \times 4$ factorial are as follows:

$$CT = \frac{Y^2_{....}}{mvkr} = 14671.845$$

$$SS_M = \frac{\Sigma Y^2_{i...}}{vkr} - CT = \frac{460.8^2 + 567.0^2}{36} - CT = 156.645$$

$$SS_V = \frac{\Sigma Y^2_{.j..}}{mkr} - CT = \frac{324.6^2 + ... + 377.7^2}{24} - CT = 77.018$$

$$SS_K = \frac{\Sigma Y^2_{..k.}}{mvr} - CT = \frac{241.8^2 + ... + 275.2^2}{18} - CT = 39.549$$

$$SS_{MV} = \frac{\Sigma Y^2_{ij..}}{kr} - CT - SS_M - SS_V$$

$$= \frac{150.6^2 + \ldots + 208.8^2}{12} - CT - SS_M - SS_V = 9.187$$

$$SS_{MK} = \frac{\Sigma\, Y^2_{i \cdot k \cdot}}{vr} - CT - SS_M - SS_V$$

$$= \frac{112.8^2 + \ldots + 159.1^2}{9} - CT - SS_M - SS_V = 24.370$$

$$SS_{VK} = \frac{\Sigma\, Y^2_{\cdot jk \cdot}}{mr} - CT - SS_V - SS_K$$

$$= \frac{76.5^2 + \ldots + 99.7^2}{6} - CT - SS_V - SS_K = 5.808$$

$$SS_{MVK} = \frac{\Sigma\, Y^2_{ijk \cdot}}{r} - CT - SS_M - SS_V - SS_K - SS_{MV} - SS_{MK}$$
$$- SS_{VK}$$

$$= \frac{36.9^2 + \ldots + 58.6^2}{r} - CT - SS_M - SS_V - SS_K - SS_{MV} -$$
$$SS_{MK} - SS_{VK}$$

$$= 7.365$$

These sums of squares plus those computed for blocks, total, and error are presented in the analysis of variance table (Table 9.13). A few comments about computation of the interaction sums of squares are advisable. Each of the first order interactions was computed from its corresponding 2-way table of totals (Table 9.12). The marginal totals in the 2-way tables supply the totals for the main effect sums of squares. The second order interaction (MVK) is computed from the totals in Table 9.11. Note that for any interaction sum of squares, that in addition to the CT we also subtract the sums of squares of any effect which is completely within that interaction (i.e., for SS_{MV} we subtracted SS_M and SS_V; and for SS_{MVK} we subtract all other treatment sums squares). In effect, each interaction sum of squares is being found as a residual sum of squares. For example, consider the 6 moisture by variety totals given at the top of Table 9.12. The differences among the 6 treatments are due to main effects as well as the interaction. To determine the interaction sum of squares we must subtract the main effect sum of squares from the sum of squares among $\left(\dfrac{\Sigma Y^2_{ij \cdot \cdot}}{kr} - CT \right)$ the 6 groups.

TABLE 9.13.   ANOVA FOR THE SEED SIZE EXPERIMENT

Source of Variation	df	SS	MS	F	Probability
Blocks	2	13.68	—	—	—
M	1	156.65	156.65	103.19	<0.01
V	2	77.02	38.51	25.37	<0.01
K	3	39.55	13.18	8.68	<0.01
MV	2	9.19	4.60	3.03	>0.05
MK	3	24.37	8.12	5.35	<0.01
VK	6	5.81	0.97	<1	>0.05
MVK	6	7.37	1.23	<1	>0.05
Error	46	69.83	1.518	—	—
Total	71	403.47	—	—	—

Now let us examine the ANOVA table and determine what inferences can be drawn about the data. As in the case of the 2-factor experiment we must proceed from the highest order interaction to the main effects. Since the second order interaction (MVK) effect is nonsignificant, we may now examine the first order interaction effects. We note that MV and VK interaction effects are nonsignificant, while the moisture level by potassium level interaction is significant. It is the nature of this interaction which we must now consider.

The means for the 8 moisture by potassium treatments are plotted in Fig. 9.3. It can be seen that the interaction of moisture and potassium is simply a greater rate of increase in seed size with increasing potassium levels for adequate moisture than for the deficient level of moisture. In fact increasing levels of potassium resulted in little or no change in seed size for deficient moisture levels. To examine the figure in a slightly different way, consider the differences between the 2 moisture levels at a given level of potassium. The interaction can be seen as an increasing difference between means for the moisture levels with increasing levels of potassium.

If the researcher wishes, some additional analyses may be performed. For example, assuming that potassium levels are quantitative and equally spaced, orthogonal polynomials (Appendix Table F.1) could be used to further examine the potassium response curves. Once it has been determined what degree of polynomial equations would best describe the data, then a regression equation can be computed for the observed data. A figure could be drawn which presents the best fit regression lines as opposed to Fig. 9.3, which is simply a line graph plotted through the treatment means.

Many researchers are tempted to apply mean comparison procedures after a significant interaction. Such an approach should be discouraged. Since differences between means may be due to either main effects and/or interaction, such a test generally provides little insight into the nature of interaction. Also, when quantitative treatment levels (i.e., levels of potassium) are used, comparisons between levels are of little importance, since the actual treatment levels are somewhat arbitrary.

Once a treatment has been included in a significant interaction, specific

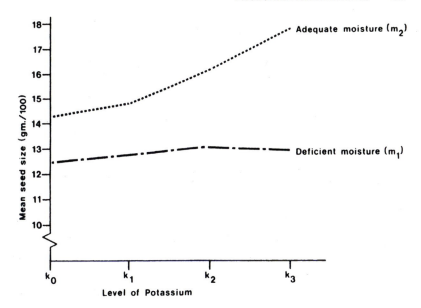

FIG. 9.3. RESPONSE LINE THROUGH TREATMENT MEANS FOR SIGNIFICANT PO-
TASSIUM BY MOISTURE INTERACTION

recommendations about the main effect are not possible. In fact, in many
cases any statements regarding the main effects would be misleading. The
present example illustrates the problem. The test of hypothesis for moisture
and potassium main effects indicates significant effects. With regard to
moisture it would be appropriate to conclude that adequate moisture resulted in greater seed size than deficient moisture conditions. It would not
be appropriate to state any specific amount by which seed size is greater,
since the amount depends on the level of potassium. At best the researcher
could state a range ($+1.9$ to $+4.8$ g/100 seeds) of superiority for the adequate
moisture level for the potassium levels studied. This in effect is describing
the interaction, not the main effect. For the potassium main effect it would
simply be incorrect to claim that the highest level (k_3) results in increased
seed size since that is clearly not the case for the deficient moisture level.

For the present example, the only main effect that can be meaningfully
interpreted is that for variety. Examination of the variety means ($v_1 = 13.5$,
$v_2 = 13.6$, $v_3 = 15.7$; SEM $= (MS_E/24)^{1/2} = 0.25$), and standard error of
means indicate that v_1 and v_2 are essentially equal, while seed size for v_3 is
greater than for the first 2 varieties.

In effect, the variety comparisons are being made with 24 replications
($m \cdot k \cdot r$), since these varieties were found to be independent of moisture and
potassium level. That is, the third variety can be expected to be greater in

seed size by the same amount, while the other two varieties are equal regardless of the moisture and potassium treatment. That is not to say that the variety means would have the same magnitude, but rather the same relative relationship.

Two other factorial experiments which are common in agriculture and food science research are the split-plot and repeated measures designs. The split-plot experiment occurs when levels of the second treatment factor are randomized within the levels of the first factor, rather than combinations of both factors being randomized simultaneously.

The repeated measures experiment results from the same experimental unit being measured more than once during the experimental period. If time effects are to be considered in the analysis, then the repeated measures analysis should be used. For details of these analyses, the reader is referred to Steel and Torrie (1960).

PROBLEMS

9.1. A study was conducted to determine the effect of microwave cooking and thawing time on the ATP concentration of meat. The experiment contained 9 treatments (3 × 3 factorial), in a completely random design. Complete the factorial analysis of variance and interpret the results.

Microwave Cooking (Sec/g Tissue)	Thawing Time (min)	ATP (μmol/g Tissue) Replication			
		1	2	3	4
0	0	2.5	2.3	2.2	2.6
	30	1.5	1.5	1.9	1.4
	60	1.2	1.4	0.9	1.1
1	0	2.6	2.7	2.5	2.6
	30	1.8	1.3	1.9	1.8
	60	1.5	1.7	1.6	1.7
2	0	1.7	1.8	1.9	1.8
	30	1.6	1.8	1.6	1.5
	60	1.6	1.7	1.4	1.4

9.2. A researcher was interested in the effectiveness of grass buffer strips in reducing fecal coliform loading rates from sludge-treated plots. Water samples and volume of runoff were measured at 0, 5, and 10 ft (1 ft = 0.305 m) distance into the grass buffer strip from the control and sludge-treated plots. The experiment was designed as an RCB. Analyze the data and summarize the results.

Treatment	Buffer Strip (ft)	Fecal Coliform (Log Numbers) Block				
		1	2	3	4	5
Control	0	7.32	6.61	7.56	6.95	6.75
	5	6.72	6.68	7.52	6.72	6.70
	10	7.03	6.51	6.98	6.50	6.02
	20	6.65	6.03	7.41	6.36	5.76
Sludge	0	9.93	8.05	10.47	9.84	7.61
	5	9.05	7.11	9.64	8.77	7.41
	10	8.05	6.71	8.94	7.88	7.06
	20	7.58	6.54	8.27	7.31	6.54

9.3. An experiment was conducted with 2 strains of rats (6/treatment) subjected to 1 of 2 treatments for either 5 or 20 days. Examine the resulting analysis of variance and table of means and summarize the results.

ANOVA

Source	Probability
Strain (S)	<0.01
Treatment (T)	<0.01
Days (D)	<0.05
ST	>0.05
SD	>0.05
TD	<0.01
STD	>0.05
Error Mean Square	4200

Strain	Treatment	Days	Mean Serum Glucose (mg/100 ml)
A	1	5	272
A	1	20	223
A	2	5	354
A	2	20	342
B	1	5	316
B	1	20	251
B	2	5	386
B	2	20	394

REFERENCES

For additional examples of the analysis and interpretation of factoria experiments see:

COCHRAN, W.G. and COX, C.M. 1957. Experimental Designs, 2nd Edi tion. John Wiley & Sons, New York.

SATTERTHWAITE, F.E. 1946. An approximate distribution of estimates o variance components. Biometrics 2, 110.

SNEDECOR, G.W. and COCHRAN, W.G. 1967. Statistical Methods, 6tl Edition. Iowa State Univ. Press, Ames.

STEEL, R.G.D. and TORRIE, J.H. 1960. Principles and Procedures of Sta tistics. McGraw-Hill Book Company, New York.

YATES, F. 1937. The design and analysis of factorial experiments. Com monw. Bur. Soil Sci. Tech. Commun. 35.

Examples of factorial applications in recent literature:

DAVIS, D.R., COCKRELL, C.W. and WIESE, K.F. 1980. Can pitting ii green beans: relation to vacuum, pH, nitrate, phosphate, copper, and iroi content. J. Food Sci. 45, 1411.

FIELD, R.A., SANCHEZ, L.R., KUNSMAN, J.E. and KRUGGEL, W.G. 198C Heme pigment content of bovine hemopoietic marrow and muscle. J. Foo Sci. 45, 1109.

ICE, J.R., HAMANN, D.D. and PURCELL, A.E. 1980. Effects of pH, en zymes, and storage time on the rheology of sweet potato puree. J. Foo Sci. 45, 1614.

MILLS, E.W., PLIMPTON, R.F. and OCKERMAN, H.W. 1980. Residual ni trite and total microbial plate counts of hams as influenced by tumblin, and four ongoing nitrite levels. J. Food Sci. 45, 1297.

10

Analysis of Covariance

In Chapters 7, 8, and 9 we studied the analysis of variance technique. Chapter 8 in particular dealt with analyses for designs (RCB and Latin squares) which help control experimental variation (error). When the blocking is used to control for a possible source of variation, it is referred to as a design technique for local control. Frequently sources of variation are encountered which are difficult or impossible to control by blocking. The analysis of covariance (ANOCOVA) deals with one or more added variables, referred to as covariates, used to measure variability in the uncontrolled variables. The analysis of covariance makes use of both the techniques of regression and analysis of variance to account for the variation associated with the covariate.

The presentation of the ANOCOVA at this point in the text relies largely on a graphic and intuitive approach, followed by an example. Since regression is an inherent part of the ANOCOVA, the reader may wish to refer to Chapters 12 and 13, which introduce linear regression techniques.

USES OF COVARIANCE ANALYSIS

The importance of the ANOCOVA technique in data analysis is twofold:

(1) To increase precision by removing from the experimental error any variation in the dependent variable associated with the covariate.
(2) To adjust the treatment means of the dependent variable for differences existing in the covariate.

Let us consider each aspect of the ANOCOVA using the data of Table 10.1 as an example. The data consist of initial and final weights on 5 replicates (blocks) of rabbits. It should be obvious to the reader that considerable variation exists in the initial weight of the rabbits. Since initial weights are taken before application of the treatment, the researcher is interested in (1) determining how much of the variation in final weight (if any) can be attributed to variation in initial weight, and (2) in what way and

TABLE 10.1. INITIAL (X) AND FINAL (Y) WEIGHTS OF RABBITS IN KG ON TREATMENTS

Block	Treatment A X_{ij}	A Y_{ij}	B X_{ij}	B Y_{ij}	Totals $X_{.j}$	$Y_{.}$
1	0.39	1.31	0.51	1.37	0.90	2.6
2	0.64	1.73	0.69	1.58	1.33	3.3
3	0.41	1.12	0.75	1.64	1.16	2.7
4	0.33	1.26	0.48	0.96	0.81	2.2
5	0.53	1.63	0.57	1.25	1.10	2.8
Totals $(X_{i.}, Y_{i.})$	2.30	7.05	3.00	6.80	X.. = 5.30	Y.. = 13.8

to what degree did the differences in initial weight of rabbits affect trea ment differences in final weight.

In Fig. 10.1 we have plotted the data of treatment A to illustrate th concept of error control. The observed values are represented by dots ($\cdot$). linear regression line has been constructed through the data to describe th relationship between initial and final weight. The technique for determir ing the regression line is not important at this time, but will be presente later. A vertical line has been constructed at $\bar{X}$ as a reference line. I addition a dotted line has been drawn from each data point to the vertic line parallel to the regression line.

Note the variation in final weights (1.12 to 1.73). The question is ho much of this variation was due to variation in initial weight? Since th regression line has a slope, final weight is to some degree associated wit initial weight. Ideally, the researcher should have had rabbits of identica initial weight. Since this was not the case, suppose that we adjusted th final weight of each rabbit to a constant initial weight, say $\bar{X}$. According t the regression line, rabbits with smaller initial weights had smaller fina weights, and greater initial weights resulted in greater final weights.

Graphically the adjusted final weight of a rabbit is at the intersection of and a line parallel to the regression line constructed through the observe final weight. From Fig. 10.1 we should observe that the adjusted weight fo the first rabbit (0.39, 1.31) is ~1.49 kg. The second rabbit (0.64, 1.73), wit a large initial weight, will result in an adjusted final weight of ~1.27 kg

It is of interest to note the effect of the adjustment on the variability final weight. For the observed Y values, the observations range from 1.12 t 1.73 kg; while the adjusted Y values range from 1.25 to 1.60. The reductio in variability is a direct result of removing the variation associated wit differences in initial weight. A similar graph could be constructed fo treatment B.

The second important use of analysis of covariance is the adjustment c treatment means ($\bar{Y}$) for differences between covariate means ($\bar{X}$). Dif ferences between adjusted treatment means may be greater or smaller tha the observed differences. The direction and magnitude of the adjustmen for each treatment mean depends on the regression line and on the re lationship between the covariate mean ($\bar{X}$) and the grand mean ($\bar{X}..$).

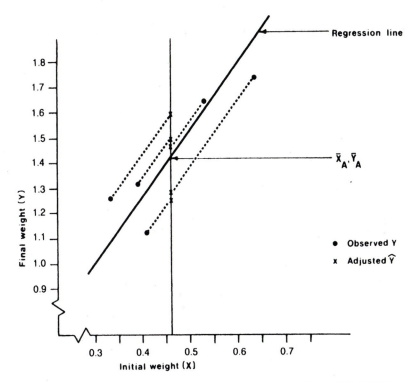

FIG. 10.1.   ILLUSTRATION OF ADJUSTMENT BY REGRESSION OF THE OBSERVA-
TIONS OF TREATMENT A

Figure 10.2 is a graphic presentation of the adjustment of treatment means. Parallel regression lines have been constructed for each treatment through the treatment means $(\bar{X}, \bar{Y})$. Since ideally the researcher would like to compare treatments free of the effect of the covariate, the process is to adjust treatment means to a common X value (for example, $\bar{X}. .$).

For treatment A when $\bar{X}_A$ was less than $\bar{X}. .$, the effect is a greater adjusted mean $(\hat{Y}_A)$ than the observed mean. Adjustment of the mean of treatment B resulted in a reduction of the treatment mean. The observed difference between the treatment means $(\bar{Y}_A - \bar{Y}_B)$ is 0.05 kg; after adjustment the difference $(\hat{Y}_A - \hat{Y}_B)$ is approximately 0.41 kg. The interpretation is that, had the researcher been able to conduct the experiment with rabbits of a uniform initial weight, we would expect the difference between treatments to be 0.41 kg.

THE LINEAR ADDITIVE MODEL

The linear additive model for an analysis of covariance simply involves the addition of a regression term to the model. For example, for an RCB

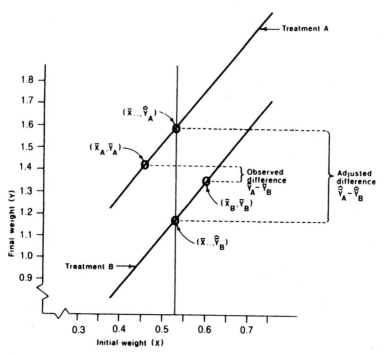

FIG. 10.2. ILLUSTRATION OF ADJUSTMENT BY ANALYSIS OF COVARIANCE OF TREATMENT MEANS

design the covariance model would be

$$Y_{ij} = \mu + \rho_i + \tau_j + \beta(X_{ij} - \bar{X}..) + \varepsilon_{ij}$$

where β is the true slope of the linear regression line

X_{ij} is the observed value of the covariate corresponding to the Y_{ij} observation

$\bar{X}..$ is the grand mean of the covariate and the other terms in the model are the same as presented in Chapter 8.

For a completely random design, with a 2-factor, factorial arrangement of treatments, the model would be:

$$Y_{ijk} = \mu + \alpha_i + \gamma_j + (\alpha\gamma)_{ij} + \beta(X_{ijk} - \bar{X}...) + \varepsilon_{ijk}$$

As might be expected, the assumptions of the ANOCOVA are essentially those of both ANOVA and linear regression. That is, the ε's are NID $(0, \sigma^2)$, the X's are fixed, and the regression is independent of block and treatment effects. The fact that X's are seldom fixed in ANOCOVA is not a serious problem. If the assumption that the X's are not affected by treatment is violated, the researcher must cautiously interpret the analysis. That is necessary since the treatment means have been adjusted to some extent for treatment effects.

TESTING ADJUSTED TREATMENT MEANS

Presented in Table 10.2 is the format and the notation for the ANO-COVA. Note that a number of rows and columns have been added compared to the ANOVA table (i.e., a regression line, sum of squares of X, cross products, etc.). Although it is not presented in Table 10.2, the adjusted sum of squares for blocks could be computed in the same manner. That is, a "Block + Error" line and a line for "between adjusted blocks" could be added to the table. For completeness we have included the "Regression" line, although it is omitted in many texts. The ANOCOVA table as presented is primarily a format to assist with computations. The abbreviated table that is normally presented includes only the "Adjusted SS," MS, F ratios, and probabilities for hypotheses of interest.

Let us complete the ANOCOVA for the rabbit data of Table 10.1. The equations for the various sums of squares and cross products are presented in Table 10.3. The equations for the sum of squares of X and Y are the same as presented in Chapter 8. Only the sum of cross products (xy) represents a new concept. The equations for the sum of cross product terms differ from the sum of squares in that they involve multiplying X by Y, rather than the squaring of items.

The quantities computed in Table 10.3 have been organized in an ANO-COVA in Table 10.4. Although it has little value here, we have included an "Adjusted Block" to reinforce the concept of adjustment. The process is to add the source of interest to the error to form a new line. Then compute the Adjusted Treatment SS as Adjusted Treatment + Error SS − Adjusted Error SS. If, for example, a factorial experiment were being analyzed by ANOCOVA, each main effect and interaction source of variation would be adjusted in this manner.

The shortened version of the analysis of covariance is presented in Table 10.5. The novice statistician might fail to recognize this table as an ANO-COVA containing adjusted effects. The essential difference is the presence of the regression source of variation, although this might in practice just as well have been labelled initial weight. To avoid confusion the writer should include the linear model as presented earlier in the chapter.

Again, let us examine the two functions of the ANOCOVA in relation to our example. First with respect to error control, we need to compare the experimental variance before and after adjustment. The MS_E before adjustment is $E_{xx}/df_E = 0.0648$. The adjusted mean square error ($S^2_{y \cdot x}$) is 0.0214. However, these two numbers should not be compared directly since the error associated with the estimate of the slope must also be taken into account. The appropriate estimate of experimental variances for the ANO-COVA is

TABLE 10.2. SYMBOLIC ANOCOVA FOR RCB DESIGN

Source of Variation	df	Sums of Squares and Products			Adjusted	
		xx	xy	yy	df	SS
Blocks	$r-1$	R_{xx}	R_{xy}	R_{yy}		
Treatments	$t-1$	T_{xx}	T_{xy}	T_{yy}		
Error	$(r-1)(t-1)$	E_{xx}	E_{xy}	E_{yy}	$(r-1)(t-1)-1$	$E_{yy} - (E_{xy})^2/E_{xx}$
Regression					1	$(E_{xy})^2/E_{xx}$
Treatment + Error	$r(t-1)$	S_{xx}	S_{xy}	S_{yy}	$r(t-1)-1$	$S_{yy} - (S_{xy})^2/S_{xx}$
Adjusted treatments					$t-1$	$SS_{T+E} - SS_E$

$$b = \frac{E_{xy}}{E_{xx}}$$

TABLE 10.3. COMPUTATIONAL EQUATIONS FOR ANOCOVA SUMS OF SQUARES AND PRODUCTS (DATA FROM TABLE 10.1)

Correction terms:

$$CT_X = X^2../rt = 2.8090$$
$$CT_Y = Y^2../rt = 19.1823$$

$$CT_{XY} = \frac{(X..)(Y..)}{rt} = 7.3405$$

Total

$$\Sigma x^2 = \Sigma X^2_{ij} - CT_X = 0.1646$$
$$\Sigma y^2 = \Sigma Y^2_{ij} - CT_Y = 0.5726$$
$$\Sigma xy = \Sigma(X_{ij})(Y_{ij}) - CT_{XY} = (0.39)(1.31) + \ldots + (0.57)(0.125) - CT_{XY} = 0.2087$$

Block

$$R_{xx} = \Sigma X^2_j/t - CT_X = 0.0863$$
$$R_{yy} = \Sigma Y^2_j/t - CT_Y = 0.3072$$

$$R_{xy} = \frac{\Sigma (X_j)(Y_j)}{t} - CT_{XY} = \frac{(0.90)(2.68) + \ldots + (1.10)(2.88)}{2} - CT_{XY}$$

$$= 0.1506$$

Treatment

$$T_{xx} = \Sigma X^2_i/r - CT_X = 0.0490$$
$$T_{yy} = \Sigma Y^2_i/r - CT_Y = 0.0062$$

$$T_{xy} = \frac{\Sigma(X_i)(Y_i)}{r} - CT_{XY} = \frac{(2.30)(7.05) + (3.00)(6.80)}{5} - CT_{XY}$$

$$= -0.0175$$

Error terms are computed by difference:

$$Error = Total - Block - Treatment$$
$$E_{xx} = 0.1646 - 0.0863 - 0.0490 = 0.0293$$
$$E_{yy} = 0.2592$$
$$E_{xy} = 0.0756$$

Treatment + Error terms are a summation:

$$S_{xx} = T_{xx} + E_{xx} = 0.0873$$
$$S_{yy} = T_{yy} + E_{yy} = 0.2654$$
$$S_{xy} = T_{xy} + E_{xy} = 0.0581$$

Equations for other quantities are given in Table 10.2.

TABLE 10.4.   ANOCOVA FOR RABBIT DATA OF TABLE 10.1

Source	df	Sums of Squares and Products			Adjusted	
		xx	xy	yy	df	ss
Blocks	4	0.0863	0.1506	0.3072		
Treatments	1	0.0490	−0.0175	0.0062		
Error	4	0.0293	0.0756	0.2592	3	0.0641
Regression					1	0.1951
Treatment + Error	5	0.0783	0.0581	0.2654	4	0.2223
Adjusted treatment					1	0.1582
Block + Error	8	0.1156	0.2262	0.5664	7	0.1238
Adjusted block					4	0.0597

TABLE 10.5   ABBREVIATED ANOCOVA FOR RABBIT DATA

Source	df	SS	MS	F
Blocks	4	0.0597	0.0149	< 1
Treatment	1	0.1582	0.1582	7.39
Regression	1	0.1951	0.1951	9.12
Error	3	0.0641	0.0214	—

$$S_{y \cdot x}^2 \left[1 + \frac{T_{xx}}{(t - 1)E_{xx}} \right]$$

and is referred to as the effective error mean square. For our example, we would compute

$$0.0214 \left[1 + \frac{0.0490}{(2 - 1)(0.0293)} \right] = 0.0572$$

The relative precision of the two techniques is 0.0648/0.0572 = 1.13. In other words, it would require 113 replicates without covariance to be as effective as 100 replicates with covariance. One additional factor should be considered in evaluating the effectiveness of the two procedures, that is, the loss of 1 degree of freedom from the covariance MS_E. Only when the degrees of freedom are small (as in this case) do we need to be concerned. For this particular example, there is little advantage to ANOCOVA with respect to error control.

The second function of ANOCOVA is concerned with the process of adjusting treatment means. To correctly adjust the observed means the slope of the regression line of Fig. 10.2 must be computed.

$$\text{Slope} = b = \frac{E_{xy}}{E_{xx}} = 2.58$$

This says that for every 1 kg increase in initial weight, final weight increased 2.58 kg. The adjusted treatment means ($\hat{Y}_{i.}$) are commonly adjusted to the mean of the covariate ($\bar{X}_{..}$). The equation for adjustment of treatment means is

$$\hat{Y}_i = \bar{Y}_i - b(\bar{X}_{i.} - \bar{X}_{..})$$

Computing the means and solving the equation, we find:

$$\hat{Y}_A = 1.41 - 2.58(0.46 - 0.53) = 1.59$$

$$\hat{Y}_B = 1.36 - 2.58(0.60 - 0.53) = 1.18$$

As shown in Fig. 10.2, the difference between the adjusted treatment means is greater than that between the observed means. The researcher should not assume that this is generally the case. The adjustment of means may result in an increase, decrease, or no change in the difference between means.

It is important that the researcher keep in mind both functions of the ANOCOVA. The benefits from either error control or adjustment of means may merit the use of ANOCOVA even when there is no advantage with respect to the other function.

To further study the adjusted treatment means, the researcher would generally be interested in determining the standard error of an adjusted treatment mean ($S_{\hat{Y}_i}$).

$$S_{\hat{Y}_i} = S_{y \cdot x} \left[\frac{1}{r} + \frac{(\bar{X}_{i.} - \bar{X}_{..})^2}{E_{xx}} \right]^{1/2}$$

For our example

$$S_{\hat{Y}_A} = 0.1463 \left[\frac{1}{5} + \frac{(0.46 - 0.53)^2}{0.0293} \right]^{1/2}$$

$$= 0.0887$$

Since $t = 2$ and $n_A = n_B$, $S_{\hat{Y}_A} = S_{\hat{Y}_B} = 0.0887$.

This standard error equation will result in an individual standard error for each treatment. The magnitude of the standard error depends on $S^2_{Y \cdot X}$ and the amount of adjustment required ($\bar{X}_{i.} - \bar{X}_{..}$), which differs from treatment to treatment. In cases where treatment differences are small for the covariate, an average standard error has been suggested

$$S_{\hat{Y}_i} = S_{Y \cdot X} \left[\frac{1}{r} + \frac{T_{xx}}{(t-1)(E_{xx})} \right]^{1/2}$$

Since treatment differences are relatively large for our example, the indi
vidual SE is preferred.

Researchers frequently employ data transformations as an alternative t
the ANOCOVA to remove the effect of the X variable. The ones mos
commonly used have one of the following general forms, $Y - X$, $\dfrac{Y-X}{X}$, or Y/X

In some cases these or similar transformations will produce satisfactor
results. However, in most cases they are simply incorrect. In each case, th
general form implies that a one unit change in X should result in a one uni
change in Y, and that for ratios, the regression should pass through th
origin.

This was clearly not the case for our example, and the use of these form
would clearly have been inappropriate. The computed relationship showe
that the adjustment required more than 1 kg (2.58 kg) change in fina
weight, when initial weight changed by 1 kg.

The advantages of using the ANOCOVA are frequently overlooked b
researchers. Often the covariate data is relatively inexpensive compared t
the difficulties and expense incurred when trying to obtain more uniformit
of the experimental units by selection or design. To many, the analysis i
poorly understood and to some appears to involve data manipulation. How
ever, ANOCOVA is a very valuable technique when the assumptions can b
satisfied and error control by design is either impossible or too expensive

PROBLEMS

10.1.   The researcher of problem 9.2 observed that the degree of groun
cover (grass density) was not uniform from plot to plot. A rating scal
was devised to evaluate the degree of ground cover to be used as
covariate. With the data below and that of 9.2, complete the analysi
of covariance. Comment on the effectiveness of the ANOCOVA.

Treatment	Buffer Strip	Ground Cover Rating Block				
		1	2	3	4	!
Control	0	9	7	9	8	'
	5	8	6	10	7	!
	10	9	8	8	9	(
	20	10	9	10	8	'
Sludge	0	8	9	10	9	(
	5	7	8	9	8	!
	10	9	7	8	9	!
	20	8	9	9	7	!

REFERENCES

COCHRAN, W.G.   1957.   Analysis of covariance; its nature and uses. Biometrics *13*, 261.

OSTLE, B. and MENSING, R.W.   1975.   Statistics in Research.   Iowa State Univ. Press, Ames.

SMITH, H.F.   1957.   Interpretation of adjusted treatment means and regressions in analysis of covariance. Biometrics *13*, 282.

STEEL, R.G.D. and TORRIE, J.H.   1960.   Principles and Procedures of Statistics.   McGraw-Hill Book Company, New York.

Examples of analysis of covariance application in recent literature:

CRANDALL, P.G. and KESTERSON, J.W.   1980.   BOD and COD determinations on citrus waste strains and component parts.   J. Food Sci. *45*, 134.

FORTIN, A., SIMPFENDORFER, S., REID, J.T., AYOLA, H.J., ANRIQUE, R. and KERTZ, A.F.   1980.   Effect of level of energy intake and influence of breed and sex on the chemical composition of cattle.   J. Anim. Sci. *51*, 604.

NEELY, J.D., JOHNSON, B.H. and ROBISON, O.W.   1980.   Heterosis in estimates for measures of reproductive traits in crossbred boars.   J. Anim. Sci. *51*, 1070.

SCHONBACHER, B.D. and CROUSE, J.D.   1980.   Growth and performance of growing-finishing lambs exposed to long or short photoperiods.   J. Anim. Sci. *51*, 943.

11

Chi-square

INFERENCES ABOUT σ^2

In earlier chapters we learned that confidence limits for μ could be constructed for samples from normally distributed populations (or approximately normal) using the t-distribution. At that time little was said concerning inferences about σ^2. We will now introduce the χ^2 distribution which will allow us to make confidence interval statements about σ^2 based on samples from normal distributions.

The distribution of the sum of squares $[\Sigma(Y - \bar{Y})^2]$ of n, normal and independent quantities (Y) divided by σ^2 is known as Chi-square (χ^2), with n − 1 degrees of freedom. That is

$$\chi^2 = \frac{\Sigma(Y_i - \bar{Y})^2}{\sigma^2} = \frac{(n - 1)S^2}{\sigma^2}, \text{ with } n - 1 \text{ df}$$

We can construct confidence intervals for σ^2 from the equation using Appendix Table H.1. The following statement would be true with $1 - \alpha$ probability.

$$P\left(\chi^2_{1 - \alpha/2 \ (n - 1)} < \chi^2 < \chi^2_{\alpha/2 \ (n - 1)}\right) = 1 - \alpha$$

Substituting the χ^2 equation into the probability statement, we have

$$P\left(\chi^2_{1 - \alpha/2 \ (n - 1)} < \frac{(n - 1)S^2}{\sigma^2} < \chi^2_{\alpha/2 \ (n - 1)}\right) = 1 - \alpha$$

With a few algebraic manipulations, we find a confidence interval statement about σ^2.

$$P\left(\frac{(n - 1)S^2}{\chi^2_{\alpha/2(n - 1)}} < \sigma^2 < \frac{(n - 1)S^2}{\chi^2_{1 - \alpha/2(n - 1)}}\right) = 1 - \alpha$$

Let's return to the example in Chapter 4 concerning the water content of Susan's Sophisticated Sausages. Seven random samples resulted in a mean percentage water content of 31.0 with a variance of 3.00. Suppose we wish to compute the 95% confidence limits for the σ^2 in H_2O content.

$$P \left(\frac{6(3)}{\chi^2_{0.025(6)}} < \sigma^2 < \frac{6(3)}{\chi^2_{0.975(6)}} \right) = 1 - 0.05$$

$$P \left(\frac{18}{14.4} < \sigma^2 < \frac{18}{1.24} \right) = 0.95$$

$$L_1 = \frac{18}{14.4} = 1.25$$

$$L_2 = \frac{18}{1.24} = 14.5$$

Therefore, we can conclude that we are 95% confident that the interval from 1.25 to 14.5 includes the true population variance.

Frequently the researcher may be more interested in confidence statements concerning the population standard deviation (σ). If so, the researcher may simply compute the square root of the limits for variance, that is, the 95% confidence limits for σ are:

$$L_1 = \left(\frac{(n - 1)S^2}{\chi^2_{\alpha/2(n - 1)}} \right)^{1/2}$$

$$= 1.25^{1/2} = 1.12$$

$$L_2 = \left(\frac{(n - 1)S^2}{\chi^2_{1 - \alpha/2(n - 1)}} \right)^{1/2}$$

$$= 14.5^{1/2} = 3.81$$

GOODNESS OF FIT TEST

Many variables in the food sciences and agriculture are the result of classifying observations into various categories, for example, the number of fertile vs nonfertile animals, the eye colors of a number of individuals, or resistance vs nonresistance to a certain disease. The test statistic for data such as these is the χ^2 statistic defined as follows:

$$\chi^2 = \Sigma \left[\frac{(\text{observed} - \text{expected})^2}{\text{expected}} \right]$$

$$= \ \Sigma \left[\frac{(f_i - \hat{f}_i)^2}{\hat{f}_i} \right] \text{with (k - 1) df, where k} $$
equals the number of classes.

Note that the observed minus expected is squared. Therefore, departures from the expected in either direction result in positive χ^2 values. Thus rejection of a null hypothesis of no difference would be indicated by large values of χ^2. Even though rejection is indicated by only one tail (large values), deviations in either direction are tested since the negative deviation (and the positive) is positive when squared.

Now consider a couple of examples to illustrate the application of the χ^2 goodness of fit test. Suppose we wish to determine if the sex ratio of captured gray squirrels is 1:1. An experiment is conducted in which 57 squirrels are identified by sex.

Hypotheses:

$$H_o: P_\delta \ = \ P_\varphi$$
$$H_1: P_\delta \ \neq \ P_\varphi$$

Test statistic:

$$f_\delta \ = \ 20, \ f_\varphi \ = \ 37$$
$$\hat{f}_\delta \ = \ nP_\delta \ = \ 57 \ (0.5) \ = \ 28.50$$
$$\hat{f}_\varphi \ = \ nP_\varphi \ = \ 57 \ (0.5) \ = \ 28.50$$

$$\chi^2 \ = \ \Sigma \ \frac{(f_i - \hat{f}_i)^2}{\hat{f}_i}$$

$$= \ \frac{(20 - 28.50)^2}{28.50} \ + \ \frac{(37 - 28.50)^2}{28.50} \ = \ 5.07 \text{ with 1 df}$$

Probability of the test statistic:

From Appendix Table H.1 we find:

$$\chi^2_{0.025(1)} \ = \ 5.02$$
$$\chi^2_{0.010(1)} \ = \ 6.63$$

Given that $\alpha \ = \ 0.05$, we would reject H_o since the probability is $0.01 < P < 0.025$. Therefore the researcher would conclude that the sex ratio of captured squirrels is not 1:1.

A plant geneticist conducted a linkage experiment in which the absence of linkage was indicated by the expected 9:3:3:1 ratio for the 4 phenotypes. The geneticist classified 306 plants and computed the following:

Hypothesis:
    H_o: The true ratio is 9:3:3:1.
    H_1: The true ratio is not 9:3:3:1.

Test statistic:

$$f_1 = 161, \quad f_2 = 58, \quad f_3 = 63, \quad f_4 = 24$$

$$\hat{f}_1 = 306 \left(\frac{9}{16} \right) = 172.125$$

$$\hat{f}_2 = \hat{f}_3 = 306 \left(\frac{3}{16} \right) = 57.375$$

$$\hat{f}_4 = 19.125$$

$$\chi^2 = \Sigma \; \frac{(f_i - \hat{f}_i)^2}{\hat{f}_i}$$

$$= \frac{(161 - 172.125)^2}{172.125} + \; \cdots \; + \frac{(24 - 19.125)^2}{19.125}$$

$$= 2.520, \text{ with 3 df}$$

Probability of the total statistic:

$$\chi^2_{0.50(3)} = 2.37$$
$$\chi^2_{0.25(3)} = 4.11$$

If $\alpha = 0.05$, then accept H_o, since $0.25 < P < 0.50$. Thus the geneticist can conclude that the results are not inconsistent with the 9:3:3:1 ratio for the absence of linkage.

CONTINGENCY TABLES

The application of χ^2 to categorical data like that of the previous section is sometimes referred to as 1-way classification. Discrete data can frequently be classified in 2 ways, leading to what is referred to as r × c tables. The r and the c refer to the number of rows and columns in the 2-way table. Such data can frequently be analyzed by the χ^2 test known as a contingency table analysis. The null hypothesis is that rows and columns are independent. That is, the results observed for the row classification are not associated with the results observed for the column classification and vice versa. The simplest of the r × c contingency table analysis is for the 2 × 2.

Suppose that the presence or absence of certain internal parasites had also been recorded for the squirrel data of the previous section, resulting in the data of Table 11.1. The expected values for contingency tables given the null hypothesis of independence of rows and columns are computed as $\hat{f}_{ij} = \frac{R_i C_j}{n}$; where $\hat{f}_{ij}$ is the expected value for the ith row and jth column,

TABLE 11.1. CLASSIFICATION OF TRAPPED GRAY SQUIRRELS BY SEX AND INTERNAL PARASITISM
2 × 2 Contingency Table.

Sex	Parasites		Total
	Present	Absent	
Male	14	6	20
Female	18	19	37
Total	32	25	57

TABLE 11.2. EXPECTED VALUES ($\hat{f}$) FOR DATA OF TABLE 11.1

Sex	Parasites		Total
	Present	Absent	
Male	11.23	8.77	20.0
Female	20.77	16.23	37.0
Total	32.0	25.0	57.0

R_i and C_j are the totals for the ith row and the jth column, respectively, and n is the total sample size. Thus the expected value for parasitized males is f_{11} = (20)(32)/57 = 11.23. Expected values for the data of Table 11.1 are presented in Table 11.2, and would be analyzed as follows:

Hypotheses:

H_o: Sex is independent of parasitism.
H_1: Sex is not independent of parasitism.

Test statistic:

$$\chi^2 = \frac{(14 - 11.23)^2}{11.23} + \frac{(18 - 20.77)^2}{20.77} + \frac{(6 - 8.77)^2}{8.77} + \frac{(19 - 16.23)^2}{16.23}$$

$$= 2.400, \text{ with } (r - 1)(c - 1) \text{ df or 1 df}$$

Probability of test statistic:

$$\chi^2_{0.25(1)} = 1.32$$
$$\chi^2_{0.10(1)} = 2.71$$

Given $\alpha = 0.05$, we would accept H_o since $0.10 < P < 0.25$. Therefore, we can conclude that the evidence does not indicate a differential rate of parasitism for the male and female gray squirrel population.

Suppose that physical therapists wished to compare 3 physical therapy regimes. Thirty subjects were assigned to each regime and after a specified period of time their progress was rated (Table 11.3). Note that since all

TABLE 11.3.   PROGRESS RATINGS UNDER 3 PHYSICAL THERAPY REGIMES
r × c Contingency Table.

Regime	Progress Rating				Total
	−	0	+	+ +	
I	4	11	9	6	30
II	0	4	15	11	30
III	2	10	15	3	30
Total	6	25	39	20	90

regimes were assigned the same number of subjects, the expected values will be identically the same in each column. For the first $(-)$ progress rating, the expected value $(\hat{f}_{i1})$ is $\dfrac{(30)(6)}{90} = 2.00$. Since the row totals are the same for all regimes, expected values will be the same within each column.

The χ^2 test statistic is not in fact an exact test, but rather an approximation. For the approximation to be reasonably good, no expected value (f) should be too small. The reason for requiring a large f is that the χ^2 approximation assumes that observed frequencies are normally distributed about the expected values. As the expected value becomes smaller and smaller, the distribution of observed values must necessarily become more and more skewed. The question is then how large should we require $\hat{f}$ to be? A commonly stated rule is that no expected value should be less than 5. Experience has led many statisticians to believe that this rule is too strict. Perhaps a more moderate rule which would provide adequate approximation by χ^2 is that all $\hat{f}$ should be greater than 1, and not more than 20% of the $\hat{f}$ should be less than 5. If the data fail to meet this requirement, then 1 of 3 actions must be taken to correct the problem or the χ^2 test should not be applied.

(1)   Collect more data.
(2)   Combine adjacent classifications if the pooled classes make sense.
(3)   Drop the column(s) or row(s) violating the rule.

Note that the application of either of the last 2 steps would result in loss of degrees of freedom and some decrease in sensitivity, and in the latter case data would be discarded.

Since in our example, 25% of the $\hat{f}$ are less than 5, we will combine the − and 0 classifications to increase the expected values. The pooled data, expected values, and the computed χ^2 value are given in Table 11.4. The probability of the computed χ^2 is less than 0.025, leading to the conclusion that the progress rating is dependent on the physical therapy regime. The logical question would be to ask which regimes are different.

An exact procedure has not been agreed on by all statisticians. A procedure which might be useful for interpretive purposes would be to continue to apply r × c contingency table analyses to identify which rows or columns are or are not different. Such a procedure in general leads to a number

TABLE 11.4.   POOLED PROGRESS RATINGS FROM TABLE 11.3

Regime	Progress Rating			Total
	− and 0	+	+ +	
I	15	9	6	30
II	4	15	11	30
III	12	15	3	30
Total	31	39	20	90
$\hat{f}_{ij}$	10.33	13.00	6.67	30

$$\chi^2 = 13.004, \text{ with 4 df}$$
$$P < 0.025$$

of nonindependent tests, and the user should be cautioned that the probabil ities are not exact.

With this word of caution, let's proceed to examine the data. From inspec tion of Table 11.4, it would appear that perhaps regimes I and II are equal while III is different. This would suggest that we conduct a 2 × 3 contin gency analysis to determine if regimes I and III are different. This analysi: is presented in Table 11.5. The computed χ^2 supports the observation tha regimes I and III did not have different progress ratings. At this point the evidence supports the conclusion that regime II progress ratings wer: different, while I and III were equal. One additional confirming analysi: could be run. That is, a 2 × 3 analysis with regime I and III pooled agains regime II. Although this analysis hardly seems necessary, the reader ma: wish to complete it for his own satisfaction.

TABLE 11.5.   COMPARISON OF REGIMES I AND III

Regime	Progress Rating			Total
	− and 0	+	+ +	
I	15	9	6	30
III	12	15	3	30
Total	27	24	9	60
$\hat{f}_{ij}$	13.5	12.0	4.5	30

$$\chi^2 = 2.833, \text{ with 2 df}$$
$$P > 0.05$$

PROBLEMS

11.1.   A survey concerning the Equal Rights Amendment (ERA) was con ducted with the following results. Is there evidence that support o ERA is dependent on voters' age? Complete the contingency tabl: analysis and write an interpretation.

Age	ERA	
---	For	Against
< 35	43	37
35–60	108	111
> 60	66	41

11.2. An experiment was conducted to determine if trained and untrained taste panelists evaluated a certain food product the same. Complete the contingency table analysis.

Panel	\multicolumn{5}{c}{Preference Rating}
---	0	1	2	3	4
Trained	2	7	9	15	7
Untrained	0	1	4	23	12

11.3. Two insecticides were compared for their effectiveness in controlling a certain plant insect. Complete the contingency table analysis.

	Insecticide	
---	X	Y
Effective	14	16
Ineffective	4	9

REFERENCES

Additional examples of contingency table analysis may be found in:

MENDENHALL, W. and OTT, L. 1976.   Understanding Statistics, 2nd Edition.   Durbury Press, North Scituate, Mass.

ZAR, J.H. 1974.   Biostatistical Analysis.   Prentice-Hall, Englewood Cliffs, N.J.

Examples of chi-square applications in recent literature:

BOWDEN, D.M. 1980.   Feed utilization for calf production in the first lactation by 2-year-old F_1 crossbred beef cows.   J. Anim. Sci. *51*, 304.

BUTTIN, H.C. and TREVINO, J.E. 1980.   Acceptability of microwave and conventionally baked potatoes.   J. Food Sci. *45*, 1425.

SEGERSON, E.C. and GONAPATHY, S.N. 1980.   Fertilization of ova in selenium/vitamin E-treated ewes maintained on two planes of nutrition.   J. Anim. Sci. *51*, 386.

Part III

Statistics in Business and Research

12

Linear Regression

Regression analysis is an extremely versatile tool. However, like all statistical techniques, it is subject to misuse and abuse. This chapter examines simple linear regression. Although this is a somewhat limited tool, a clear discussion of it will facilitate an understanding of related techniques.

A SIMPLE EXAMPLE

Suppose that we are concerned with the relationship between cooking time and meat tenderness. Table 12.1 shows the recorded observations on 7 samples. An examination of the values shown in Table 12.1 indicates that as cooking time increases, shear force decreases (i.e., the meat becomes more tender). Plotting this information on a simple graph in Fig. 12.1 shows this relationship more clearly.

TABLE 12.1. OBSERVATIONS ON COOKING TIME AND SHEAR FORCE FOR MEAT SAMPLES

Observation No.	Cooking Time (min) X	Shear Force (lb) Y	(Newtons) Y
1	10	500	2224
2	9	550	2446
3	11	425	1890
4	8	550	2446
5	10	450	2002
6	12	350	1557
7	7	625	2780

Two questions come readily to mind:

(1) Is the apparent relationship real or just a chance result?
(2) Can we define a line that represents the average relationship between time and tenderness?

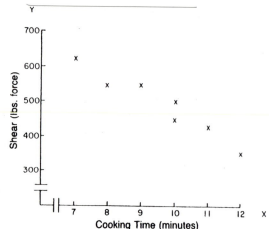

FIG. 12.1.   SHEAR FORCE
AND COOKING TIME FOR
MEAT SAMPLES

1 lb = 4.448 newtons.

Simple linear regression uses the information available to define a line
that represents the average relationship. Regression analysis also permits
us to test whether or not the relationship is real or just a chance result.

In order to draw a line on a graph, we need to know the Y intercept and the
slope of the line. Figure 12.2 illustrates the graphing of a line with the
equation $Y = a + bX$. Regression analysis uses all of the observations
available to determine both a and b for the equation $Y = a + bX$. The
technique used is called least-squares because it minimizes the sum of
squared deviations about the line.

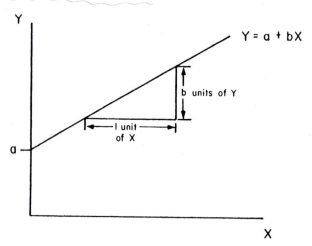

FIG. 12.2.   STRAIGHT LINE GRAPH ILLUSTRATING INTERCEPT
AND SLOPE

Figure 12.3 shows a typical regression line illustrating the deviations about the line. The regression line has several characteristics:

(1) The line passes through the point $\bar{x}$, $\bar{y}$ (i.e., the means of X and Y).
(2) The sum of the deviations about the line is zero.
(3) The sum of the deviations squared will be a minimum (i.e., any other line that could be drawn would have a greater sum of squared deviations).

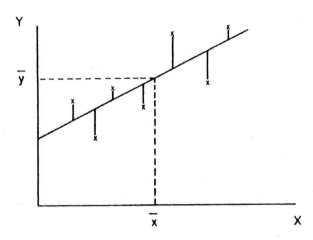

FIG. 12.3.   REGRESSION LINE WHICH MINIMIZES THE SUM OF SQUARED DEVIATIONS

The formulas needed to calculate a regression line are:

$$b = \frac{\Sigma XY - \dfrac{(\Sigma X)(\Sigma Y)}{n}}{\Sigma X^2 - \dfrac{(\Sigma X)^2}{n}}$$

$$a = \bar{y} - b\bar{X}$$

Table 12.2 shows our original observations and the calculations needed to use our formulas. In our case:

$$\Sigma XY = 32{,}100$$
$$\Sigma X = 67$$
$$\Sigma Y = 3{,}450$$
$$\Sigma X^2 = 659$$
$$n = 7$$

TABLE 12.2. CALCULATIONS NEEDED TO ESTIMATE A SIMPLE LINEAR REGRESSION

	Cooking Time X	X^2	Shear Y	Y^2	XY
	10	100	500	250,000	5000
	9	81	550	302,500	4950
	11	121	425	180,625	4675
	8	64	550	302,500	4400
	10	100	450	202,500	4500
	12	144	350	122,500	4200
	7	49	625	390,625	4375
Sum	67	659	3450	1,751,250	32,100
Mean	9.571		492.857		

Table 12.3 shows how these numbers are used to develop the equation:

$$\hat{Y} = 990.712 - 52.017X$$

This is the line of "best" fit in the sense that it minimizes the sum of squared deviations or error sum of squares (ESS).

TABLE 12.3. CALCULATION OF REGRESSION SLOPE AND INTERCEPT

$$b = \frac{\Sigma XY - \dfrac{(\Sigma X)(\Sigma Y)}{n}}{\Sigma X^2 - \dfrac{(\Sigma X)^2}{n}}$$

$$= \frac{32,100 - \dfrac{(67)(3450)}{7}}{659 - \dfrac{(67)^2}{7}}$$

$$= \frac{32,100 - 33,021.429}{659 - 641.286}$$

$$= \frac{-921.429}{17.714}$$

$$= -52.017$$

$$a = \bar{y} - b\bar{x}$$

$$a = 492.857 - (-52.017)(9.571)$$

$$a = 990.712$$

Table 12.4 shows the calculated value of Y (i.e., $\hat{Y}$; read Y-hat) for each X, the Y deviation (i.e., $Y - \hat{Y}$), and the deviation squared. It should be noted that the sum of deviations is nearly zero. The difference is due to rounding errors. The sum of squared deviations is 2963.714.

TABLE 12.4. THE ESTIMATED VALUE OF Y AND. DEVIATIONS ABOUT THE REGRESSION LINE

	X	Y	$\hat{Y}$	$Y - \hat{Y}$	$(Y - \hat{Y})^2$
	10	500	470.542	29.458	867.774
	9	550	522.559	27.441	753.008
	11	425	418.525	6.475	41.926
	8	550	574.576	−24.576	. 603.980
	10	450	470.542	−20.542	421.974
	12	350	366.508	−16.508	272.514
	7	625	626.593	− 1.593	2.538
Sum	67	3450		0.155	2963.714

In practice we would not calculate the error sum of squares directly as was done in Table 12.4. Instead, we should use the formula: Error Sum of Squares = Total Sum of Squares − Regression Sum of Squares

or

$$\text{Error Sum of Squares} = \left[\Sigma Y^2 - \frac{(\Sigma Y)^2}{n} \right] - \left[b^2 \left(\Sigma X^2 - \frac{(\Sigma X)^2}{n} \right) \right]$$

In our case:

$$\text{ESS} = \left[1,751,250 - \frac{(3450)^2}{7} \right] - \left[(-52.017)^2 \left(659 - \frac{(67)^2}{7} \right) \right]$$

$$= [1,751,250 - 1,700,357.143] - [2705.768\ (17.714)]$$

$$= 50,892.857 - 47,929.974$$

$$= 2962.883$$

The error sum of squares represents the unexplained variation. That is, it represents variation in Y (meat tenderness) not explained by variation in X (cooking time). The greater the unexplained variation, the less reliable are our results. If unexplained variation is zero, we can predict perfectly. In other words, if unexplained variation is zero, then all of the observations lie exactly on our regression line.

We can use these formulas to develop a line of "best fit" without any understanding of the underlying statistical assumptions. In fact, the preceding material was simply an arithmetic procedure. It could be used whether or not any statistical analysis was desired. However, by making a few basic assumptions, this regression line procedure can be used to develop inferential statements of considerable value.

ASSUMPTIONS OF REGRESSION ANALYSIS

The most basic assumption underlying regression analysis is that there is a unilateral causal relationship between X and Y. That is, variations in X result in variations in Y. But variations in Y do not result in variations in X. Another way of saying the same thing is that the X's are fixed and predetermined; while the Y's are observations on a random variable whose mean is a function of X. It is common practice to assume that Y is normally distributed. Figure 12.4 shows this concept where each point on the regression line represents the mean of a population of Y's. The populations are identical except for their means. The assumption of normality is necessary in order to test hypotheses.

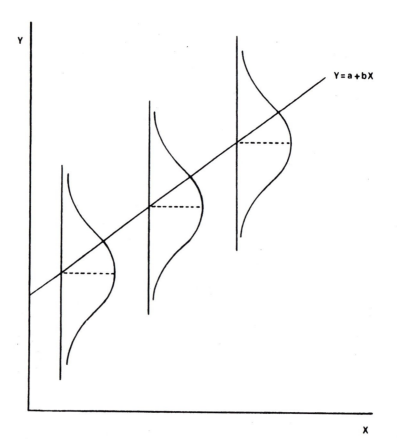

FIG. 12.4.   REGRESSION LINE SHOWING THE NORMAL DISTRIBUTION OF Y ABOUT EACH POINT ON THE LINE

INTERPRETING THE REGRESSION RESULTS

Ordinarily, a researcher estimates a regression line in order to estimate a causal relationship that is believed to exist on *a priori* grounds. The anticipated cause and effect relationship is based on the experience and judgment of the researcher; regression analysis merely provides an acceptable arithmetic procedure for quantifying the hypothesized relationship. The interpretation of the estimated slope must be made within the context of the problem studied. What statistics offers is supporting evidence of the reliability of these statements within a probability framework.

In regression analysis, the true model is defined as:

$$Y = \alpha + \beta X + \epsilon$$

where α = the true but unknown intercept
β = the true but unknown slope
and ϵ = the true but unobservable deviation between the true regression line and the observed Y

As is generally true in statistics, Greek letters are used to denote the parameters or true values, while English letters are used to denote statistics or estimates of true values, so that

α is estimated by a
β is estimated by b
and
ϵ is estimated by e

The error sum of squares (ESS) is an estimate of the innate variability of Y (i.e., variations in Y after the effect of X has been taken into account). The ESS enables us to estimate the reliability of our results and to make probability statements about future events.

The first step is to estimate the variance of Y net of the influence of X (i.e., $\sigma_{y \cdot x}^2$). Since we had n observations and estimated 2 parameters (i.e., α and β) we have $n - 2$ degrees of freedom. As a result, the variance of Y given X is estimated as follows:

$$s_{y \cdot x}^2 = \frac{\text{Error Sum of Squares}}{n - 2}$$

$$= \frac{2962.883}{5}$$

$$= 592.577$$

The standard error of Y is the square root of the variance. In our case

$$s_{y \cdot x} = \sqrt{592.577}$$

$$= 24.343$$

The standard error of Y gives an indication of the amount of variability that exists about the regression line. The smaller $s_{y \cdot x}$, the better the line fits the data.

Another, more readily understandable measure of "goodness of fit" is the coefficient of determination (r^2) which calculates the percentage of variation in Y explained by variation in X. The formula is as follows:

$$r^2 = \frac{\text{Regression Sum of Squares}}{\text{Total Sum of Squares}}$$

or

$$r^2 = \frac{\text{Total Sum of Squares} - \text{Error Sum of Squares}}{\text{Total Sum of Squares}}$$

The second statement shows clearly that as the error sum of squares approaches zero, r^2 approaches the value of 1.0. Conversely, if the regression explains none of the variation in Y, r^2 will approach the value of zero.

In our case

$$r^2 = \frac{47,929.974}{50,892.857}$$

$$= 0.942$$

Thus, we would say that in our case, 94.2% of the variation in Y is explained by variation in X. This is quite good. We would be reluctant to make decisions with a regression line that had an r^2 of less than 0.80. If we cannot explain more than 80% of the variation, we should attempt to develop more explanatory variables.

In order to test whether or not the relationship defined by our regression line is real or just a chance occurrence, we must construct two new statistics.

The slope of the line (b) calculated in Table 12.3 is an estimate of the true slope (β). The true slope may be equal to zero (i.e., there is no relationship between X and Y). Since b is an estimate of a true population parameter, it has a distribution. The variance of b is a function of the variance of Y because b is calculated as a function of Y.

The formula for the variance of b is

$$s_b^2 = \frac{s_{y \cdot x}^2}{\Sigma X^2 - \dfrac{(\Sigma X)^2}{n}}$$

$$s_{y \cdot x}^2 = \frac{\text{error Sum of Squares}}{n-2}$$

$$= \frac{592.577}{17.714}$$

$$= 33.452$$

The standard deviation of b is the square root of the variance of b.

$$s_b = \sqrt{33.452}$$

$$= 5.784$$

We can use the standard deviation of b (s_b) to test our estimate of the slope of the regression line (b).

The test statistic used for this purpose is Student's t. The formula for this test statistic is:

$$t = \frac{b - \beta}{s_b}$$

where   b =   estimate of the regression slope

β =   the true but unknown slope

s_b =   the standard deviation of b

For our particular problem, we want to know whether or not the slope of the regression line is significantly different from zero. In standard form, we would write that our null hypothesis (H_o) is that the true slope (β) is equal to zero.

$$H_o: \beta = 0$$

Our alternative hypothesis (H_1) is that the true slope (β) is not equal to zero.

$$H_1: \beta \neq 0$$

We calculate t using the assumption that the null hypothesis is correct. As a result, the formula is as follows:

$$t = \frac{b - \beta}{s_b}$$

where   b =   −52.017

β =   0 (by assumption)

s_b =   5.784

so that $\quad t \; = \; \dfrac{-52.017 \; - \; 0}{5.784}$

$\qquad\quad = \; -8.993$

We compare our calculated value of t with the tabular value for Student's t in order to determine whether or not this answer could be expected on the basis of chance.

Using Student's t for an error rate of 0.05 with 5 degrees of freedom (i.e., $n - 2$), the tabular value is 2.571. If our calculated value of t is greater (in absolute terms) than the tabular value of t, we reject the null hypothesis. That is, we conclude that what we have observed was not a "chance occurrence," but rather that the slope of the true regression line differs significantly from zero.

Figure 12.5 shows graphically the test that we have just made. The bell-shaped curve shows the distribution of events that we would expect with the t test. If the null hypothesis is true and we perform the test over and over again, 95% of our calculated values should lie between ± 2.571. Since this is a two-tailed test, very large numbers and very small (i.e., negative) numbers would indicate a difference from zero. In our case, the calculated value of −8.993 lies well outside the acceptance region.

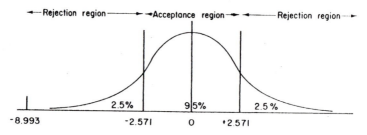

FIG. 12.5.   GRAPHIC PRESENTATION OF THE t-TEST

An important word of caution that should be kept in mind is that regression results that are valid within the range of observations may be useless outside that range. In our case, we had observations that were developed within a range of cooking times from 7 to 12 min. Within that range, the fit of our data seemed to be very good. A high percentage of the variation in tenderness was explained; the slope of the line differed significantly from zero; and visual inspection of the data implied a linear relationship. However, for cooking times that are shorter than 7 min or longer than 12 min, we cannot be sure that we can accurately predict behavior.

Figure 12.6 shows a "True Relationship" superimposed upon our data and regression results. If we are interested in what happens at shorter cooking times, we must set up appropriate trials, record the data, and calculate the necessary statistics. Whether or not we can use a linear regression depends upon how well the regression line fits the data within the range of observations.

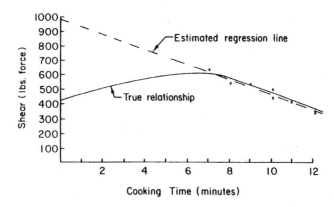

FIG. 12.6.   TRUE AND ESTIMATED RELATIONSHIPS FOR COOKING TIME AND TENDERNESS

1 lb = 4.448 newtons.

CONFIDENCE INTERVAL ABOUT THE REGRESSION LINE

Instead of simply testing whether or not the estimated regression slope is significantly different from zero, we may be interested in developing a statement about the entire regression line. Since the true regression line is unknown, our calculated regression line is only an estimate of the true regression line. In other words, since a and b are functions of Y (a random variable), they are also random variables. If we assume that Y is normally distributed, then a and b will also be normally distributed. The mean of the distribution of a's is α (the true intercept). The mean of the distribution of b's is β (the true slope). However, we normally have a single set of observations and a single calculated a and b (i.e., a single set of estimates of α and β). As a result, we want to construct statistical statements that provide an insight into the reliability of our results.

There are two issues to be addressed:

(1) What can we say about points along the regression line? That is, since points along the regression line represent estimates of population means, can we construct a confidence interval about this series of means?

(2) What can we say about predicting a single future event? Even if we knew the true regression line, we could only speak of future events in terms of probabilities due to the innate variability of the material studied. As a result, when discussing the prediction of a single future event we must account for the variation in the material studied as well as the variation in the regression line which is only an estimate of the true relationship.

Chapter 4 presented the concept of constructing a confidence interval about the mean of a population. Now we want to construct an interval about each point on the regression line as well as confidence limits concerning future events. Figure 12.7 shows a hypothetical example.

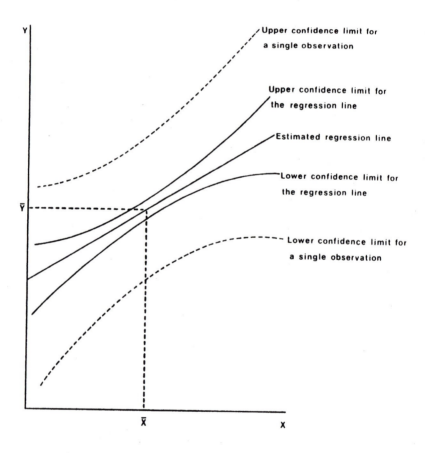

FIG. 12.7.   HYPOTHETICAL REGRESSION LINE WITH CONFIDENCE LIMITS

Figure 12.7 shows an estimated regression line. The inner curved boundaries represent the confidence interval about the regression line. The curvature of the boundary is due to the unknown slope since different slopes may be estimated as a result of chance variation. In general, the construction of a confidence interval is a time consuming task that is rarely undertaken. However, it often yields valuable information. For example, the interval about the regression line may be very small indicating that we have a reliable estimate of "the average relationship between X and Y." At the same time, we may have a very large interval associated with predicting a single future event indicating that the innate variability of the material is so great that although we are able to predict average behavior, we cannot predict a single future event for decision purposes.

The confidence interval about the regression line is calculated by the following formula:

$$CL(\hat{Y}) = \hat{Y} \pm t\, s_{y\cdot x} \left[\frac{1}{n} + \frac{(X-\bar{x})^2}{\Sigma X^2 - \frac{(\Sigma X)^2}{n}} \right]^{1/2}$$

where $\hat{Y}$ = estimated population mean given X

$\quad$ t = Student's t for the appropriate degrees of freedom and probability level

$s_{y\cdot x} \left[\dfrac{1}{n} + \dfrac{(X-\bar{x})^2}{\Sigma X^2 - \frac{(\Sigma X)^2}{n}} \right]^{1/2}$ = standard deviation of the regression line given X

For our example, we would calculate the 95% confidence interval with 5 degrees of freedom for X = 7 as follows:

$$CL(\hat{Y}) = 626.59 \pm (2.571)(24.343) \left[\frac{1}{7} + \frac{(7-9.571)^2}{17.714} \right]^{1/2}$$

$$= 626.59 \pm 62.5859 \,[0.1429 + 0.3732]^{1/2}$$

$$= 626.59 \pm 62.5859(0.7184)$$

$$= 626.59 \pm 44.96$$

We could state this as:

$$P(581.63 \leq \mu_{y\cdot x} \leq 671.55) = 0.95$$

In other words, the probability that the population mean $\mu_{y\cdot x}$ lies between 581.63 and 671.55 is 0.95. Actually, the true population mean for X =

7 either lies within the interval (P = 1.0) or it does not (P = 0). It would be more correct to say that if we repeated the experiment a large number of times and constructed a confidence interval each time, 95% of the intervals constructed would contain the true population mean. This concept was discussed earlier in Chapter 4. Figure 4.11 illustrates the repeated construction of a confidence interval about a population mean for 25 independent samples. Table 12.5 shows the raw data and the intervals calculated to graph Fig. 12.8.

When constructing the confidence limits for a single future Y the following formula is used:

$$CL(A \text{ Single } Y) = \hat{Y} \pm t \, s_{y \cdot x} \left[1 + \frac{1}{n} + \frac{(X - \bar{x})^2}{\Sigma X^2 - \frac{(\Sigma X)^2}{n}} \right]^{1/2}$$

It is easier to visualize what this formula means if it is written as

$$CL(A \text{ Single } Y) = \hat{Y} \pm t \left[s_{y \cdot x}^2 + \frac{s_{y \cdot x}^2}{n} + s_{y \cdot x}^2 \left[\frac{(X - \bar{x})^2}{\Sigma X^2 - \frac{(\Sigma X)^2}{n}} \right] \right]^{1/2}$$

where $s_{y \cdot x}^2$ reflects the innate variability of the material studied

$\dfrac{s_{y \cdot x}^2}{n}$ reflects the variability associated with a sample mean

and $s_{y \cdot x}^2 \left[\dfrac{(X - \bar{x})^2}{\Sigma X^2 - \frac{(\Sigma X)^2}{n}} \right]$ reflects the variability associated with the estimated slope.

In order to calculate the 95% confidence limits for a single future Y given that X = 7 and that there are 5 degrees of freedom, we use the formula as follows:

$$CL(A \text{ Single } Y) = 626.59 \pm (2.571)(24.343) \left[1 + \frac{1}{7} + 0.3732 \right]^{1/2}$$

$$= 626.59 \pm 62.5859[1.5161]^{1/2}$$

$$= 626.59 \pm 77.06$$

In other words, if we repeat the experiment with X = 7, we will expect to observe a value between 549.53 and 703.65. This may be too large an interval for practical use. Increasing sample size might reduce this value since 2 of the 3 components of the variance decrease as n increases. However, in predicting a single future Y, the innate variability of the material as

TABLE 12.5. CONSTRUCTION OF CONFIDENCE LIMITS FOR THE REGRESSION LINE AND FOR A SINGLE PREDICTED VALUE OF Y

Base Data Needed			Limits for the Regression Line			Limits for a Single Predicted Y		
			$s_{y \cdot x}\left[\dfrac{1}{n} + \dfrac{(X-\bar{x})^2}{\Sigma X^2 - \dfrac{(\Sigma X)^2}{n}}\right]^{1/2}$	$\hat{Y} \pm t_{0.05}\, s_{y \cdot x}\left[\dfrac{1}{n} + \dfrac{(X-\bar{x})^2}{\Sigma X^2 - \dfrac{(\Sigma X)^2}{n}}\right]^{1/2}$		$s_{y \cdot x}\left[1 + \dfrac{1}{n} + \dfrac{(X-\bar{x})^2}{\Sigma X^2 - \dfrac{(\Sigma X)^2}{n}}\right]^{1/2}$	$\hat{Y} \pm t_{0.05}\, s_{y \cdot x}\left[1 + \dfrac{1}{n} + \dfrac{(X-\bar{x})^2}{\Sigma X^2 - \dfrac{(\Sigma X)^2}{n}}\right]^{1/2}$	
X	$(X-\bar{x})^2$	$\hat{Y}$	$24.343\left[\dfrac{1}{7} + \dfrac{(X-\bar{x})^2}{17.714}\right]^{1/2}$	Lower Limit	Upper Limit		Lower Limit	Upper Limit
7	6.610	626.59	17.486	581.63	671.55	29.971	549.53	703.65
8	2.468	574.58	12.926	541.35	607.81	27.561	503.72	645.44
9	0.326	522.56	9.766	497.45	547.67	26.230	455.12	590.00
9.571	0.000	492.86	9.207	469.19	516.53	26.025	425.95	559.77
10	0.184	470.54	9.523	446.06	495.02	26.140	403.33	537.75
11	2.042	418.52	12.364	386.73	450.31	27.303	348.32	488.72
12	5.900	366.51	16.794	323.33	409.69	29.574	290.48	442.54

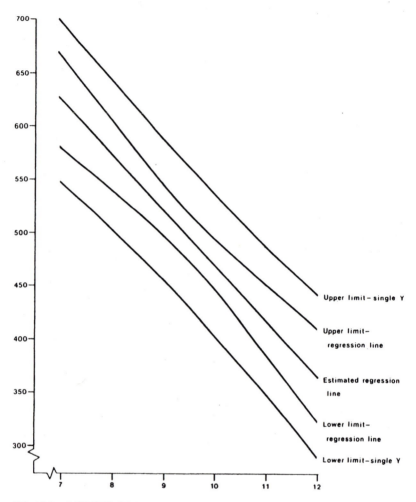

FIG. 12.8.   CONFIDENCE INTERVALS ABOUT THE REGRESSION LINE FOR COOKING
TIME—MEAT TENDERNESS

measured by $s_{y \cdot x}^2$ must always be recognized. Table 12.5 shows the raw data
and the intervals calculated to graph the confidence limits about a single
future Y shown in Fig. 12.8.

NONLINEAR DATA

A common form of nonlinearity can be overcome through a simple transformation. If the data can be plotted in a curve as shown in Fig. 12.9, an

alternative is to plot the data on a log-log paper or semilog paper (Fig. 12.10). A linear plot of the transformed data would indicate that we could use a simple linear regression to estimate the log-log or semilog relationship. Table 12.6 shows the actual (arithmetic) data and transformed (logarithmic) data.

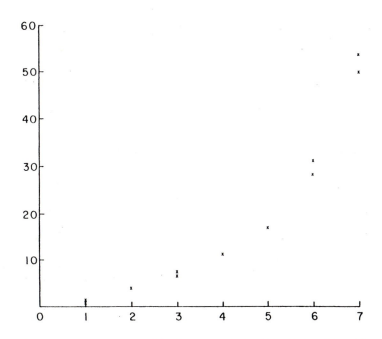

FIG.12.9. THE VALUES OF TABLE 12.6 PLOTTED ON A RECTILINEAR GRID

It should be noted that the data used are the log. of our observations on Y and the arithmetic values of X. As a result, the regression equation is of the form:

$$\log Y = a + bX$$

Tables 12.7 and 12.8 show the calculation of the intercept (a) and slope (b).

Since these calculations are in the semilog world, our regression line predicts the log of Y as shown in Table 12.9. The sum of deviations about the regression line is in terms of log. deviations. However, the important point is that by making this transformation of data, simple linear regression can handle a common form of nonlinearity.

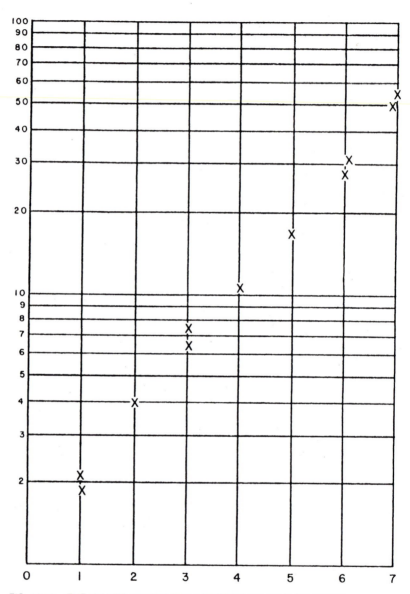

FIG. 12.10.   THE VALUES OF TABLE 12.6 PLOTTED ON SEMILOG PAPER

TABLE 12.6.   SELECTED VALUES ILLUSTRATING A NON-LINEAR RELATIONSHIP

X	Y	log. Y
1	1.9	0.2788
1	2.1	0.3222
2	4.0	0.6021
3	7.3	0.8633
3	6.5	0.8129
4	11.0	1.0414
5	17.4	1.2406
6	31.0	1.4914
6	28.5	1.4548
7	50.5	1.7033
7	54.0	1.7324

SUMMARY

This chapter has dealt with simple linear regression because a clear understanding of linear regression greatly facilitates the transition to multiple regression. In practice, it is far more common to use multiple regression because it is rare that a single explanatory variable will satisfy the needs of a researcher.

Since linear regression is a special case of multiple regression, most of the references on regression analysis are listed at the end of Chapter 13.

TABLE 12.7.   CALCULATIONS NEEDED TO ESTIMATE A SEMILOG SIMPLE LINEAR REGRESSION

	X	X^2	Y	log. Y	$(\log. Y)^2$	X log. Y
	1	1	1.9	0.2788	0.0777	0.2788
	1	1	2.1	0.3222	0.1038	0.3222
	2	4	4.0	0.6021	0.3625	1.2042
	3	9	7.3	0.8633	0.7453	2.5899
	3	9	6.5	0.8129	0.6608	2.4387
	4	16	11.0	1.0414	1.0845	4.1656
	5	25	17.4	1.2406	1.5391	6.2030
	6	36	31.0	1.4914	2.2243	8.9484
	6	36	28.5	1.4548	2.1164	8.7288
	7	49	50.5	1.7033	2.9012	11.9231
	7	49	54.0	1.7324	3.0012	12.1268
Sums	45	235	214.2	11.5432	14.8168	58.9295
Means	4.0909			1.0494		

TABLE 12.8.   CALCULATION OF SEMILOG REGRESSION SLOPE AND INTERCEPT

$$b = \frac{\Sigma XY - \dfrac{(\Sigma X)(\Sigma Y)}{n}}{\Sigma X^2 - \dfrac{(\Sigma X)^2}{n}}$$

$$= \frac{58.9295 - \dfrac{(45)(11.5432)}{11}}{235 - \dfrac{(45)^2}{11}}$$

$$= \frac{58.9295 - \dfrac{519.4440}{11}}{235 - \dfrac{2025}{11}}$$

$$= \frac{58.9295 - 47.2222}{235 - 184.0909}$$

$$= \frac{11.7073}{50.9091}$$

$$= 0.2300$$

$$a = \bar{y} - b\bar{x}$$

$$= 1.0494 - (0.2300)(4.0909)$$

$$= 0.1085$$

TABLE 12.9.   THE ESTIMATED VALUE OF Y IN A SEMILOG REGRESSION

	X	log. Y	log. $\hat{Y}$	log. Y − log. $\hat{Y}$
	1	0.2788	0.3385	−0.0597
	1	0.3222	0.3385	−0.0163
	2	0.6021	0.5685	0.0336
	3	0.8633	0.7985	0.0648
	3	0.8129	0.7985	0.0144
	4	1.0414	1.0285	0.0129
	5	1.2406	1.2585	−0.0179
	6	1.4914	1.4885	0.0029
	6	1.4548	1.4885	−0.0337
	7	1.7033	1.7185	−0.0152
	7	1.7324	1.7185	0.0139
Sums	45	11.5432	11.5435	−0.0003

PROBLEMS

12.1.   The management of Fred's Fine Fruit Stand has observed that the quantity of bananas sold varies inversely with the price. Over time the following data have been observed.

Price Cents/lb	Quantity Sold lb/Week
28	400
30	390
30	325
25	440
25	400
20	600
22	600
20	550
32	255
33	300
29	300
35	225
27	350
26	500
23	550
22	500

(a) Plot the data on a simple X, Y graph. Since the price is fixed by management, price is the causal variable (i.e., X), and consumer purchases (i.e., Y) are dependent upon price.
(b) Estimate the regression equation $Y = a + bX$.
(c) Since Fred wants to use this equation to determine his stocking policy and thereby avoid excess inventory as well as shortages, calculate and graph the confidence interval about the regression line.
(d) What are the confidence limits on sales for a given week if price is 25¢/lb, that is, the confidence limits for a single future Y?

12.2.   The following data have been observed:

X	Y
10	9
15	15
45	90
50	135
75	700

X	Y
30	48
20	25
65	300
60	200
40	85
30	33
55	130
20	20
70	410
40	62

(a) Plot the data on regular graph paper, semilog paper, and log-log paper.

(b) Select the version that provides the most nearly linear array and calculate the corresponding regression equation.

REFERENCES

Applications of linear regression in recent literature:

BLANKENSHIP, L.C., DAVIS, C.E. and MAGNER, G.J. 1980. Cooking methods for elimination of *Salmonella typhimurium* experimental surface contaminant from rare dry-roasted beef roasts. J. Food Sci. *45*, 270.

CALVERT, C.C. and SMITH, L.W. 1980. Arsenic in tissues of sheep and milk of dairy cows fed arsanilic acid and 3-nitro-4-hydroxyphenylarsonic acid. J. Anim. Sci. *51*, 414.

FORBUS, W.R., JR., SENTER, S.D., LYON, B.G. and DUPUY, H.P. 1980. Correlation of objective and subjective measurements of pecan kernel quality. J. Food Sci. *45*, 1376.

REDDY, N.R., SALUNKHE, D.K. and SHARMA, R.P. 1980. Flatulence in rats following ingestion of cooked and germinated black gram and a fermented product of black gram and rice blend. J. Food Sci. *45*, 1161.

13

Multiple Regression

Linear regression in Chapter 12 is a special case of the general multiple regression model. In multiple regression, we are dealing with a Y that is a function of several X's. If our multiple regression model reduces to a single explanatory variable, we are dealing with simple linear regression. Although the basic logic remains the same, because we are dealing with more variables, our arithmetic is more cumbersome and a few additional assumptions are necessary.

THE LINEAR ADDITIVE MODEL

In Chapter 12, we wrote the regression equation as:

$$\hat{Y} = a + bX$$

since

$$a = \bar{y} - b\bar{x}$$

we could have written the same model as

$$\hat{Y} = \bar{y} + b(X - \bar{x})$$

This alternative formulation shows more clearly the underlying logic of the linear additive model. That is, Y is equal to the general population mean (estimated by $\bar{y}$) and deviations from that mean as X deviates from its mean.

If we adopt the convention that lower case y equals the deviation from its mean (i.e., $y = Y - \bar{y}$) and that lower case x equals the deviation from its mean (i.e., $x = X - \bar{x}$), our notation becomes greatly simplified. Our model becomes:

$$\hat{Y} = \bar{y} + bx$$

where

$$b = \frac{\Sigma xy}{\Sigma x^2}$$

In a similar manner

$$r^2 = \frac{b^2 \Sigma x^2}{\Sigma y^2}$$

The true but unknown model is

$$Y = \mu + \beta x + \epsilon$$

where μ = the true population mean
β = the true slope
ϵ = the true error term

In multiple regression, we simply add additional explanatory variables.

$$Y = \mu + \beta_1 x_1 + \beta_2 x_2 + \ldots + \beta_m x_m + \epsilon$$

In linear regression, the equation defines a line where each point represents an estimated population mean (see Fig. 12.4). In multiple regression, the equation defines a plane or hyperplane where each point on the plane represents an estimated population mean.

PROPERTIES AND ASSUMPTIONS OF MULTIPLE REGRESSION

In multiple regression, the equation has the following properties:

(1) The plane defined by the equation passes through the point $\bar{y}$ at the mean of the X's (i.e., the point $\bar{x}_1, \bar{x}_2, \ldots, \bar{x}_m$).
(2) The sum of the deviations is zero (i.e., $\Sigma(Y - \hat{Y}) = 0$).
(3) The sum of squared residuals is a minimum (i.e., $\Sigma(Y - \hat{Y})^2$ is a minimum).

The basic assumptions are as follows:

(1) There is a unilateral causal relationship. Variations in the X's result in variations in Y. Variations in Y do not result in variations in X.
(2) The X's are independent of one another.
(3) The residuals (i.e., ϵ's) are random and independently distributed with mean zero and a constant variance. If tests of hypotheses are to be made, it is necessary to specify the distribution of the residuals. It is common to assume normality. As a result, this assumption is usually stated as, "The residuals are normally and independently distributed with mean zero and a constant variance: ϵ are $NID(0,\sigma^2)$."

Using these assumptions, regression analysis yields BLU estimations. BLU or Best (i.e., minimum variance) Linear (i.e., linear functions of Y) Unbiased (i.e., $E(b_i) = \beta_i$, $E(\bar{y}) = \mu$) estimators possess the desirable characteristics of being unbiased and extracting maximum information from the observed data.

STEPS IN ANALYSIS

Many classes in public speaking offer the dictum, "Before opening mouth to speak, be sure that brain is engaged." In statistics, it is also desirable to follow a similar procedure.

Theoretical Base

Statistical techniques do not replace thinking. Regression analysis cannot identify the cause and effect relationship or identify the appropriate equation form. Regression analysis merely offers an arithmetic procedure for quantifying a relationship that *has been defined by the researcher*. The first step in using regression analysis is to define the variables, their relationship to one another, and probable functional form.

If you have only a single dependent variable, you are dealing with a univariate model and are within the scope of this text. If you have two or more dependent variables simultaneously determined, you are outside the scope of this text. You are dealing with a multivariate model which is commonly referred to as "simultaneous equations." Foote (1955) presents a clear discussion of the difference between ordinary least squares and simultaneous equations. Goldberger (1964) and Johnston (1972) present the underlying mathematics as well as the arithmetic procedures required in such a circumstance.

Selecting the Equation Form

After you have divided the variables into two groups, (1) the single dependent variable and (2) the one or more independent variables, the next step is to select reasonable alternatives concerning the appropriate equation form. For example, theory may indicate that the equation must pass through the origin. Alternatively, it may be known that interaction exists. There may be a whole host of background information that can be used to select possible equation forms. On the other hand, nearly any surface can be approximated in a narrow range by a plane. As a result, a linear form may be as satisfactory as more complex forms.

A good way to start the search for an appropriate equation form is to plot the data. Admittedly, if there are more than 5 or 6 explanatory variables this can be both time consuming and misleading. The use of computers can alleviate the drudgery of plotting scatter-diagrams. However, with a large number of explanatory variables, no single X may show enough of a relationship with Y to offer a clue to the underlying relationship. As a result, the researcher may be forced to rely upon theoretical considerations, prior experience, and intuition to select the appropriate form for the equation.

THE ABBREVIATED DOOLITTLE

In practice, it is common to use a standardized computer program to calculate the regression equation. Chapter 20 illustrates the use of a typical computer routine. However, since computers are not always readily available, this section of this chapter presents an algorithm that can be used to calculate all of the statistics needed for regression analysis. Although the procedure appears to be rather formidable, it is sufficiently simple that with repeated usage the steps are often committed to memory through exposure.

The Abbreviated Doolittle Method was developed in 1878 by Doolittle while he was an engineer in the U.S. Coast and Geodetic Survey. It consists of a forward and backward solution. The forward solution enables the user to estimate the regression coefficients. The backward solution permits the user to make statistical tests.

The Forward Solution

(1) The first step is to organize the sums of squares and cross products in a table similar to that of Table 13.1. This becomes the first entry of a table that has the general format shown in Table 13.2.

(2) The forward solution proceeds by copying the first line of instruction 1 so that $A_{1j} = a_{1j}$. Then each element in that row is divided by a_{11} so that $B_{1j} = A_{1j} \div a_{11}$.

(3) For the next section, calculate the A entries as follows:
$A_{2i} = a_{2i} - A_{12}B_{1i}$.
Compute the B entries by dividing each element in the A row by A_{22}.

(4) For the last section, calculate the A entries by:
$A_{3j} = a_{3j} - (A_{13}B_{1j} + A_{23}B_{2j})$.
Compute the B entries by dividing each element in the A row by A_{33}.

This completes the forward solution. In order to handle more independent variables, the original A matrix must be enlarged but the pattern of computations is the same. By the same token, if we have only 2 independent variables, the matrix is reduced.

TABLE 13.1.   INITIAL ORGANIZATION OF REGRESSION DATA

Σx_1^2	$\Sigma x_1 x_2$	$\Sigma x_1 x_3$	$\Sigma x_1 y$
	Σx_2^2	$\Sigma x_2 x_3$	$\Sigma x_2 y$
		Σx_3^2	$\Sigma x_3 y$

TABLE 13.2.   GENERAL FORM OF THE ABBREVIATED DOOLITTLE METHOD FOR MULTIPLE REGRESSION

Instruction	A Matrix			G Matrix
	x_1	x_2	x_3	y
1 x_1	a_{11}	a_{12}	a_{13}	g_1
x_2		a_{22}	a_{23}	g_2
x_3			a_{33}	g_3
2 A_{1j}	A_{11}	A_{12}	A_{13}	A_{1y}
B_{1j}	1	B_{12}	B_{13}	B_{1y}
3 A_{2j}		A_{22}	A_{23}	A_{2y}
B_{2j}		1	B_{23}	B_{2y}
4 A_{3j}			A_{33}	A_{3y}
B_{3j}			1	B_{3y}

When the forward solution is complete, the regression coefficients and regression sum of squares may be calculated.

$$b_3 = B_{3y}$$
$$b_2 = B_{2y} - B_{23}b_2$$
$$b_1 = B_{1y} - B_{12}b_1 - B_{13}b_2$$

The Backward Solution

Table 13.3 shows the backward solution to the matrix in Table 13.2. The elements of the C matrix are calculated as follows:

(1) $C_{33} = \dfrac{1}{A_{33}}$

$C_{23} = -B_{23}C_{33}$

$C_{13} = -B_{12}C_{23} - B_{13}C_{33}$

(2) $C_{32} = C_{23}$

$C_{22} = \dfrac{1}{A_{22}} - B_{23}C_{23}$

$C_{12} = -B_{12}C_{22} - B_{13}C_{23}$

(3) $C_{31} = C_{13}$

$C_{21} = C_{12}$

$C_{11} = \dfrac{1}{A_{11}} - B_{12}C_{12} - B_{13}C_{13}$

TABLE 13.3.   THE BACKWARD SOLUTION TO A
MULTIPLE REGRESSION PROBLEM

C_{11}	C_{12}	C_{13}
C_{21}	C_{22}	C_{23}
C_{31}	C_{32}	C_{33}

Using the C matrix and the standard error of Y, we are able to calculate the standard error of each regression coefficient and thereby test for significance. The appropriate degrees of freedom are:

$$df = n - 1 - k$$

where   n   =   number of observations
        k   =   number of explanatory variables

The standard error of Y equals:

$$s_{y \cdot x} = \left(\frac{\text{Total Sum of Squares} - \text{Regression Sum of Squares}}{\text{Total Sum of Squares}} \right)^{1/2}$$

where   Total Sum of Squares $= \Sigma y^2 = \Sigma Y^2 - \dfrac{(\Sigma Y)^2}{n}$

Regression Sum of Squares $= b_1 \Sigma x_1 y + b_2 \Sigma x_2 y + b_3 \Sigma x_3 y$

The standard error of each b is:

$$s_{b_i} = \left[C_{ii} s^2_{y \cdot x} \right]^{1/2}$$

The coefficient of determination (i.e., R^2) shows the percentage of variation in Y explained by variation in X.

$$R^2 = \frac{\text{Regression Sum of Squares}}{\text{Total Sum of Squares}}$$

The use of the Abbreviated Doolittle Method is more comprehensible after it has been illustrated.

An Example with 2 Explanatory Variables

The Forward Solution.—The data in Table 13.4 represent the results of 14 trials with 7 different fertilizer levels. Experience indicates that corn yield varies with fertilizer level. In other words, corn yield is the dependent variable (Y) and fertilizer level is the independent variable (X). Figure 13.1 shows a scatter-diagram of the data listed in Table 13.4. Clearly, the relationship between X and Y is nonlinear. In this case, it appears to follow

TABLE 13.4.   OBSERVATIONS ON FERTILIZER
APPLICATIONS AND CORN YIELDS

Fertilizer Level lb N/Acre X	Corn Yield bu/Acre Y
90	85
90	83
100	90
100	92
110	95
110	94
120	100
120	101
130	100
130	99
140	95
140	93
150	88
150	90

1 lb/acre = 1.125 kg/ha.
1 bu/acre = 88 liters/ha.

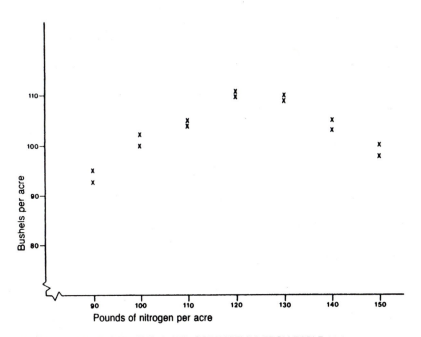

FIG. 13.1.   FERTILIZER LEVELS AND CORN YIELDS FROM TABLE 13.4

1 lb/acre = 1.125 kg/ha. 1 bu/acre = 88 liters/ha.

a quadratic relationship. As a result, we hypothesize an equation of the following form:

$$Y = a + bX - cX^2$$

If we set $X_1 = X$ and $X_2 = X^2$, we can write this as the multiple regression equation:

$$Y = a + b_1X_1 + b_2X_2$$

or

$$Y = \bar{y} + b_1(X_1 - \bar{x}_1) + b_2(X_2 - \bar{x}_2)$$

Using the data of Table 13.4 and the instructions relating to Tables 13.1 and 13.2, we can solve for the regression equation. Since it is rather time consuming to use the definitional formulas implied by Table 13.1, it is desirable to algebraically manipulate formulas dealing with deviations from means into formulas that use the original observations directly. Table 13.5 shows the general form of such a rewriting of the formulas. In general, the calculating formulas are easier to use and introduce fewer rounding errors. Table 13.6 shows the calculations necessary to complete the first part of Table 13.7 where $X_1 = X$ (i.e., the fertilizer level in lb), $X_2 = X^2$, and Y equals the yield in bu per acre.

In practice, it is desirable to keep the ratio of the largest to the smallest number in the A matrix less than or equal to 10 in order to reduce rounding errors. In our case the ratio is 623,538 (i.e., 324,240,000 ÷ 520) which by ordinary standards would be completely unacceptable. However, scaling the data to achieve a more desirable A matrix often confuses the novice statistician. Since this is a very small matrix and our first example, we will work with these data. A subsequent problem will be used to illustrate the use of scaling to modify the A matrix.

Table 13.7 shows the completion of the forward solution. With this information, we can calculate:

(1) The regression equation
(2) The coefficient of determination (R^2)
(3) The standard error of Y ($s_{y \cdot x}$)

To determine the regression equation:

$$
\begin{aligned}
b_2 &= B_{2y} \\
&= -0.0141666 \\
B_1 &= B_{1y} - B_{12}b_1 \\
&= 0.092857 - [(240)(-0.0141666)] \\
&= 0.092857 + 3.399984 \\
&= 3.492841
\end{aligned}
$$

TABLE 13.5. ALGEBRAIC MANIPULATION FROM DEFINI-
TIONAL TO CALCULATING FORMULAS

$$\Sigma x^2 = \Sigma(X - \bar{x})(X - \bar{x})$$

$$= \Sigma(X^2 - 2X\bar{x} + \bar{x}^2)$$

$$= \Sigma X^2 - 2\bar{x}\Sigma X + n\bar{x}^2$$

$$= \Sigma X^2 - 2\frac{(\Sigma X)}{n}\Sigma X + n\frac{(\Sigma X)^2}{n^2}$$

$$= \Sigma X^2 - \frac{(\Sigma X)^2}{n}$$

$$\Sigma xy = \Sigma(X - \bar{x})(Y - \bar{y})$$

$$= \Sigma(XY - X\bar{y} - Y\bar{x} + \bar{x}\bar{y})$$

$$= \Sigma XY - \bar{y}\Sigma X - \bar{x}\Sigma Y + n\bar{x}\bar{y}$$

$$= \Sigma XY - \frac{\Sigma Y}{n}\Sigma X - \frac{\Sigma X}{n}\Sigma Y + n\frac{\Sigma X}{n}\frac{\Sigma Y}{n}$$

$$= \Sigma XY - \frac{(\Sigma X)(\Sigma Y)}{n}$$

$$\Sigma y^2 = \Sigma(Y - \bar{y})(Y - \bar{y})$$

$$= \Sigma(Y^2 - 2\bar{y}Y + \bar{y}^2)$$

$$= \Sigma Y^2 - 2\bar{y}\Sigma Y + n\bar{y}^2)$$

$$= \Sigma Y^2 - 2\frac{\Sigma Y}{n}\Sigma Y + n\frac{(\Sigma Y)^2}{n^2}$$

$$= \Sigma Y^2 - \frac{(\Sigma Y)^2}{n}$$

The regression equation can be written as:

$$Y = \bar{y} + b_1x_1 + b_2x_2$$
$$Y = 93.21 + 3.493x_1 - 0.01417x_2$$

If we prefer, we can calculate the intercept using the following formula:

$$\begin{aligned}
a &= \bar{y} - b_1\bar{x}_1 - b_2\bar{x}_2 \\
&= 93.214286 - (3.492841 \cdot 120) - (-0.0141666 \cdot 14{,}800) \\
&= 93.214286 - 419.140920 + 209.665680 \\
&= -116.260954
\end{aligned}$$

This permits us to rewrite the regression equation as:

$$Y = -116.26 + 3.493X_1 - 0.01417X_2$$

TABLE 13.6. CALCULATION OF MATRIX ELEMENTS FOR TABLE 13.7

$$\Sigma x_1^2 = \Sigma X_1^2 - \frac{(\Sigma X_1)^2}{n}$$

$$= 207{,}200 - \frac{(1680)^2}{14}$$

$$= 5600$$

$$\Sigma x_1 x_2 = \Sigma X_1 X_2 - \frac{(\Sigma X_1)(\Sigma X_2)}{n}$$

$$= 26{,}208{,}000 - \frac{(1680)(207{,}200)}{14}$$

$$= 1{,}344{,}000$$

$$\Sigma x_2^2 = \Sigma X_2^2 - \frac{(\Sigma X_2)^2}{n}$$

$$= 3{,}390{,}800{,}000 - \frac{(207{,}200)^2}{14}$$

$$= 324{,}240{,}000$$

$$\Sigma x_1 y = \Sigma X_1 Y - \frac{(\Sigma X_1)(\Sigma Y)}{n}$$

$$= 157{,}120 - \frac{(1680)(1305)}{14}$$

$$= 520$$

$$\Sigma x_2 y = \Sigma X_2 Y - \frac{(\Sigma x_2)(\Sigma Y)}{n}$$

$$= 19{,}415{,}00 - \frac{(207{,}200)(1305)}{14}$$

$$= 101{,}000$$

TABLE 13.7. THE FORWARD SOLUTION TO A MULTIPLE REGRESSION PROBLEM

	\multicolumn{3}{c}{Data}		
	Σx_1^2	$\Sigma x_1 x_2$	$\Sigma x_1 y$
Instruction		Σx_2^2	$\Sigma x_2 y$
1	5600	1,344,000	520
		324,240,000	101,000
2	5600	1,344,000	520
	1	240	0.092857
3		1,680,000	-23,799.808
		1	- 0.0141666

Since X_1 equals fertilizer level and X_2 equals $X_1{}^2$, we would ordinarily write the equation as:

$$Y = -116.26 + 3.493X - 0.01417X^2$$

It should be noted that we carried 6 digits during our calculations but rounded considerably in presenting the final equation. Although it is usually necessary to carry 6 or 8 digits during calculations, there is no benefit in presenting final results with so many digits. A second item that should be noted is the negative intercept. A negative intercept in this case represents a physical impossibility since zero is obviously the smallest yield possible. However, as pointed out in Chapter 12, regression results are valid only within the range of observed data. Our smallest value of X was 90. We have no way of discussing expected yields for fertilizer applications below 90 lb of N per acre (101.25 kg/ha). The intercept is merely an arithmetic convenience. It has no valid interpretation unless we have observed data in the vicinity of X = 0, in which case, estimating the intercept may be one of our major concerns. In general, it is best to avoid trying to interpret the value of an intercept determined in regression analysis—especially when the X's are themselves some distance from the origin.

The coefficient of determination (R^2) provides a measure of how well the regression equation fits the data.

$$R^2 = \frac{\text{Regression Sum of Squares}}{\text{Total Sum of Squares}}$$

$$= \frac{b_1\Sigma x_1 y + b_2\Sigma x_2 y}{\Sigma y^2}$$

$$= \frac{(3.492841)(520) + (-0.0141666)(101,000)}{122,059 - \frac{(1305)^2}{14}}$$

$$= \frac{385.451}{414.357}$$

$$= 0.930$$

In other words, 93% of the variation in corn yield (Y) is explained by variations in nitrogen applications (X).

We can continue our analysis by testing the entire equation using the analysis of variance (ANOV) framework. Table 13.8 shows our multiple regression results summarized in an ANOV format. The sources of variation are: regression, error, and total. The degrees of freedom associated with the regression sum of squares are 2 (i.e., there are 2 explanatory variables). The total degrees of freedom are 13 (i.e., n − 1). The degrees of freedom

TABLE 13.8.   MULTIPLE REGRESSION RESULTS SUMMARIZED IN ANOV FORMAT

Sources of Variation	Degrees of Freedom	Sum of Squares	Mean Square	F
Regression	2	385.451	192.726	73.336**
Error	11	28.906	2.628	—
Total	13	414.357	—	—

**Significant at the 1% level.

associated with the error term are 11 (i.e., n − 1 - the number of explanatory variables).

The regression sum of squares and total sum of squares were calculated in the determination of R^2. The error sum of squares is calculated by subtraction. The mean squares are determined by dividing each sum of squares by its appropriate degrees of freedom. The last step is the calculation of F as the ratio of mean squares.

Using Table D.1 in the Appendix, the tabular value for F with 2 and 11 degrees of freedom is 7.20 at the 1% level. Since the calculated value of 73.336 exceeds the tabular value, we conclude that the overall regression is statistically significant. The null hypothesis is:

$$H_o: \beta_i = 0 \text{ for all } i.$$

The alternative hypothesis is:

$$H_1: \beta_i \neq 0 \text{ for at least one } i.$$

The Backward Solution.—To continue our analysis beyond this point we must calculate the backward solution to the Abbreviated Doolittle.

The backward solution is readily calculated from the A and B matrices of the forward solution.

$$C_{22} = \frac{1}{A_{22}} = \frac{1}{5600} = 0.000000595238$$

$$C_{12} = -B_{12} C_{22} = -(240)(0.000000595238)$$

$$= -0.000142857$$

$$C_{21} = C_{12} = -0.000142857$$

$$C_{11} = \frac{1}{A_{11}} - B_{12}C_{12}$$

$$= \frac{1}{5600} - (240)(-0.000142857)$$

$$= 0.000178571 + 0.0342857$$

$$= 0.0344643$$

The C matrix is the inverse of the A matrix and as such provides a check on our calculations, including rounding errors.

If we define $D_{ij} = \sum_k A_{ik} C_{kj}$, then $D_{ij} = 1$ when $i = j$, and $D_{ij} = 0$ when $i \neq j$.

In our case:

$$D_{11} = A_{11} C_{11} + A_{12} C_{21}$$

$$= (5600) (0.0344643) + (1,344,000) (-0.000142857)$$

$$= 193.000080 - 191.999808$$

$$= 1.000272$$

$$D_{12} = A_{11} C_{12} + A_{12} C_{22}$$

$$= (5600) (-0.000142857) + (1,344,000) (0.000000595238)$$

$$= 0.000001$$

$$D_{21} = A_{21} C_{11} + A_{22} C_{21}$$

$$= (1,344,000) (0.0344643) + (324,240,000) (-0.000142857)$$

$$= 0.065520$$

$$D_{22} = A_{21} C_{12} + A_{22} C_{22}$$

$$= (1,344,000) (-0.000142857) + (324,240,000) (0.000000595238)$$

$$= 1.000161$$

This check on our calculations indicates that our rounding error is unacceptably large. In practice, we would return to the original data, scale them, and redo our calculations. In this case, we will continue with these data and scale the data for the next example.

Tests of Significance

With the completion of the backward solution, individual slopes can be tested for statistical significance. Each is tested with a t-test. The general form is:

H_o: $\beta_i = 0$

H_1: $\beta_i \neq 0$

where the test statistic is:

$$t = \frac{b_i - \beta_i}{s_{b_i}}$$

and

$$s_{b_i} = \left[C_{ii}s_{y \cdot x}^2\right]^{1/2}$$

where $s_{y \cdot x}^2$ is the Error Mean Square from Table 13.8.
For our specific example:

$$b_1 = 3.493$$

$$s_{b_1} = [(0.0344643)\,(2.628)]^{1/2}$$

$$= 0.0905722^{1/2}$$

$$= 0.300952$$

$$t = \frac{3.493 - 0}{0.300952} = 11.6065$$

The tabular value for Student's t (from Appendix Table C.1) with 11 degrees of freedom is 2.201 at the 95% level of confidence. Consequently, we reject the null hypothesis that the true slope is zero. That is, the data indicate that there is a statistically significant relationship between corn yield and the level of nitrogen fertilizer applied.

In the same way, the second b value can be tested.

$$b_2 = -0.0141$$

$$s_{b_2} = \left[(0.000000595238)\,(2.628)\right]^{1/2}$$

$$= 0.000001564286^{1/2}$$

$$= 0.0012507$$

$$t = \frac{-0.0141 - 0}{0.0012507}$$

$$= -11.2737$$

Again we reject the small hypothesis that the true slope is zero.
The ANOV results shown in Table 13.8 provided a test of the overall equation. It indicated that at least one b was significantly different from zero. Using Student's t, we determined that both of our b values were significantly different from zero.

CONSTRUCTION OF A CONFIDENCE INTERVAL

When dealing with simple linear regression in Chapter 12, we constructed confidence intervals about the regression line. In multiple regression, a similar confidence interval can be constructed about the plane or hyperplane.

The confidence interval about the regression plane is calculated by the following formula:

$$CL(\hat{Y}) = \hat{Y} \pm t\, s_{\hat{y}}$$

where $\hat{Y}$ = estimated population mean given the independent variables

t = Student's t for the appropriate degrees of freedom and probability level

and $S_{\hat{y}}$ = standard deviation of the regression plane given the independent variables

For 3 variables, $S_{\hat{y}}$ is calculated as follows:

$$S_{\hat{y}} = S_{y \cdot x}^2 \left(\frac{1}{n} + C_{11}x_1^2 + C_{22}x_2^2 + C_{33}x_3^2 + 2c_{12}x_1x_2 + 2c_{13}x_1x_3 + 2c_{23}x_2c_3 \right)^{1/2}$$

Of course, with 4 or more X's, the terms within the brackets must be expanded.

In our example, the regression equation was:

$$y = -116.26 + 3.493X - 0.01417X^2$$

The 95% confidence interval about the plane where $X = 130$ is calculated as follows:

$$Y = -116.26 + 3.493(130) - 0.01417\,(16{,}900)$$

$$\doteq 98.357$$

$$CL(\hat{Y}) = 98.357 \pm t_{0.5}\, S_{\hat{y}}$$

$$t_{0.05} = 2.201$$

$$S_{\hat{y}} = 2.628 \left[\frac{1}{14} + 0.0344643\,(130 - 120)^2 + 0.000000595238 \right.$$

$$(16,900 - 14,800)^2 + 2(-0.000142857)\,(130 - 120)$$

$$\left. (16,900 - 14,800) \right]^{1/2}$$

$$S_{\hat{y}} = 2.628(0.0714286 + 3.44643 + 2.625000 - 5.999994)^{1/2}$$

$$S_{\hat{y}} = 0.375448^{1/2}$$

$$= 0.612738$$

$$CL(\hat{Y}) = 98.357 \pm 2.201\,(0.612738)$$

$$= 98.357 \pm 1.349$$

We could state this as:

$$P(97.008 \leq \mu_{y \cdot x} \leq 99.706) = 0.95$$

If we were interested in constructing the confidence limits for a single future Y, it would be necessary to add a 1 within the brackets. For this example

$$CL(A\ Single\ Y) = 98.357 \pm 2.201 \cdot 2.628\,(1 + 0.0714286 + 3.44643 +$$
$$2.625000 - 5.999994)^{1/2}$$

$$= 98.357 \pm 2.201\,(1.733046)$$

$$= 98.357 \pm 3.814$$

As with simple linear regression, the confidence interval associated with a single future value of Y is much larger than that associated with the regression plane since it must account for the innate variability of the material studied—in this case, corn yield.

INTERACTION

The concept of interaction was presented in Chapter 9 within the framework of an analysis of variance model. The regression model can also handle data where interaction is a consideration. The previous example estimated the relationship between corn yield and applications of nitrogen fertilizer. However, if phosphorus fertilizer is also used, previous experience indicates the presence of an NP interaction. Table 13.9 reports the results of 42 observations on corn yields using 3 levels of phosphorus and 7 levels of nitrogen. The middle phosphorus column is the data previously reported in Table 13.4.

TABLE 13.9.   CORN YIELDS IN BU/ACRE FOR SELECTED LEVELS OF NITROGEN AND PHOSPHORUS APPLICATIONS

Nitrogen Level lb/Acre	60 lb/Acre	Phosphorus Level 80 lb/Acre	100 lb/Acre
150	79	88	127
150	80	90	125
140	86	95	120
140	85	93	117
130	92	100	110
130	90	99	109
120	92	100	100
120	90	101	103
110	87	95	90
110	85	94	88
100	80	90	80
100	82	92	83
90	75	85	70
90	73	83	72

1 lb/acre = 1.125 kg/ha.
1 bu/acre = 88 liters/ha.

When only nitrogen was considered as an explanatory variable, the regression model was:

$$Y = a + b_1N + b_2N^2$$

In this case, an appropriate model might be:

$$Y = a + b_1N + b_2N^2 + b_3P + b_4P^2 + b_5(NP)$$

where the last term represents the interaction.

In the previous example, with only 2 independent variables, we did not scale the data. In this example, the data will be scaled. Table 13.10 presents the raw data from Table 13.9 organized as 42 observations on a single Y and 5 independent variables. The choice of scale is somewhat arbitrary but the purpose of scaling is to avoid large differences in the values contained in the A matrix.

The sums from Table 13.10 are:

$$\Sigma X_1 = 50.40$$
$$\Sigma X_2 = 62.16$$
$$\Sigma X_3 = 33.60$$
$$\Sigma X_4 = 28.00$$
$$\Sigma X_5 = 40.32$$
$$\Sigma Y = 38.75$$

TABLE 13.10.   THE DATA OF TABLE 13.9 SCALED AND ORGANIZED FOR A MULTIPLE REGRESSION MODEL

Y Yield in Hundreds of bu	X_1 Nitrogen in Hundreds of lb	X_2 X_1^2 (10,000s)	X_3 Phosphorus in Hundreds of lb	X_4 X_3^2 (10,000s)	X_5 NP Interaction $X_1 \cdot X_3$ (10,000s)
0.79	1.50	2.25	0.60	0.36	0.90
0.80	1.50	2.25	0.60	0.36	0.90
0.86	1.40	1.96	0.60	0.36	0.84
0.85	1.40	1.96	0.60	0.36	0.84
0.92	1.30	1.69	0.60	0.36	0.78
0.90	1.30	1.69	0.60	0.36	0.78
0.92	1.20	1.44	0.60	0.36	0.72
0.90	1.20	1.44	0.60	0.36	0.72
0.87	1.10	1.21	0.60	0.36	0.66
0.85	1.10	1.21	0.60	0.36	0.66
0.80	1.00	1.00	0.60	0.36	0.60
0.82	1.00	1.00	0.60	0.36	0.60
0.75	0.90	0.81	0.60	0.36	0.54
0.73	0.90	0.81	0.60	0.36	0.54
0.88	1.50	2.25	0.80	0.64	1.20
0.90	1.50	2.25	0.80	0.64	1.20
0.95	1.40	1.96	0.80	0.64	1.12
0.93	1.40	1.96	0.80	0.64	1.12
1.00	1.30	1.69	0.80	0.64	1.04
0.99	1.30	1.69	0.80	0.64	1.04
1.00	1.20	1.44	0.80	0.64	0.96
1.01	1.20	1.44	0.80	0.64	0.96
0.95	1.10	1.21	0.80	0.64	0.88
0.94	1.10	1.21	0.80	0.64	0.88
0.90	1.00	1.00	0.80	0.64	0.80
0.92	1.00	1.00	0.80	0.64	0.80
0.85	0.90	0.81	0.80	0.64	0.72
0.83	0.90	0.81	0.80	0.64	0.72
1.27	1.50	2.25	1.00	1.00	1.50
1.25	1.50	2.25	1.00	1.00	1.50
1.20	1.40	1.96	1.00	1.00	1.40
1.17	1.40	1.96	1.00	1.00	1.40
1.10	1.30	1.69	1.00	1.00	1.30
1.09	1.30	1.69	1.00	1.00	1.30
1.00	1.20	1.44	1.00	1.00	1.20
1.03	1.20	1.44	1.00	1.00	1.20
0.90	1.10	1.21	1.00	1.00	1.10
0.88	1.10	1.21	1.00	1.00	1.10
0.80	1.00	1.00	1.00	1.00	1.00
0.83	1.00	1.00	1.00	1.00	1.00
0.70	0.90	0.81	1.00	1.00	0.90
0.72	0.90	0.81	1.00	1.00	0.90

1 bu = 35.2 liters.
1 lb = 0.45 kg.

The sums of squares and cross products are shown in Table 13.11.

Table 13.12 shows the forward solution to the abbreviated Doolittle where the A and G matrices contain the corrected sums of squares and cross products. For example, the first term is:

$$\Sigma x_1^2 = \Sigma X_1^2 - \frac{(\Sigma X_1)^2}{n}$$

$$= 62.16 - \frac{(50.4)^2}{42}$$

$$= 62.16 - 60.48$$

$$= 1.68$$

The remaining values were calculated in a similar manner. Since the data have been scaled, all of the values are less than 10.

Nitrogen and phosphorus are orthogonal variables. That is, at each level of 1 variable, all levels of the other variable are represented. As a result, the corrected cross products between nitrogen and phosphorus are zero. This is a basic principle of experimental design which cannot always be achieved in practice.

Table 13.13 shows the calculation of the b values from the forward solution of Table 13.12.

TABLE 13.11. SUMS OF SQUARES AND CROSS PRODUCTS FROM THE DATA IN TABLE 13.10

Item	X_1	X_2	X_3	X_4	X_5	Y
X_1	62.1600	78.6240	40.3200	33.6000	49.7280	47.1320
X_2		101.7240	49.7280	41.4400	62.8992	58.8142
X_3			28.0000	24.1920	33.6000	31.4360
X_4				21.5488	29.0304	26.5256
X_5					41.4400	38.3204
Y						36.4971

TABLE 13.12. THE FORWARD SOLUTION FOR NITROGEN-PHOSPHORUS REGRESSION ANALYSIS

Instruction	Item	X_1	X_2	X_3	X_4	X_5	Y
	X_1	1.68	4.032	0.0	0.0	1.344	0.632
	X_2		9.7272	0.0	0.0	3.2256	1.4642
1	X_3			1.12	1.792	1.344	0.436
	X_4				2.882133	2.1504	0.692267
	X_5					2.7328	1.1024
2	A_{1j}	1.68	4.032	0.0	0.0	1.344	0.632
	B_{1j}	1.	2.4	0.0	0.0	0.8	0.376190
3	A_{2j}		0.0504	0.0	0.0	0.0	-0.0525981
	B_{2j}		1.	0.0	0.0	0.0	-1.043613
4	A_{3j}			1.12	1.792	1.344	0.436
	B_{3j}			1.	1.6	1.2	0.389286
5	A_{4j}				0.014933	0.0	-0.00533351
	B_{4j}				1.	0.0	-0.357163
6	A_{5j}					0.0448	0.091601
	B_{5j}					1.	2.0446652

TABLE 13.13. CALCULATION OF B VALUES FROM THE FORWARD SOLUTION OF TABLE 13.12

$$b_5 = B_{5v} = 2.0446652$$

$$b_4 = B_{4v} - B_{45}b_5$$
$$= -0.357163 - (0.0 \times 2.0446652)$$
$$= -0.357163$$

$$b_3 = B_{3v} - B_{34}b_4 - B_{35}b_5$$
$$= 0.389286 - (1.6 \times -0.357163) - (1.2 \times 2.0446652)$$
$$= -1.492851$$

$$b_2 = B_{2v} - B_{23}b_3 - B_{24}b_4 - B_{25}b_5$$
$$= -1.043613 - 0 - 0 - 0$$
$$= -1.043613$$

$$b_1 = B_{1v} - B_{12}b_2 - B_{13}b_3 - B_{14}b_4 - B_{15}b_5$$
$$= 0.376190 - (2.4 \times -1.043613) - 0 - 0 - (0.8 \times 2.0446652)$$
$$= 1.245129$$

With the completion of the forward solution and the knowledge that $\bar{y} = 0.922619$, the regression equation can be written as:

$$Y = 0.922619 + 1.245129x_1 - 1.043613x_2 - 1.492851x_3 - 0.357163x_4 + 2.044665x_5$$

However, it is usually more convenient to calculate the intercept and rewrite the equation.

$$a = \bar{Y} - b_1\bar{x}_1 - b_2\bar{x}_2 - b_3\bar{x}_3 - b_4\bar{x}_4 - b_5\bar{x}_5$$

$$= 0.442522$$

$$Y = 0.442522 + 1.245129X_1 - 1.043613X_2 - 1.492851X_3 - 0.357163X_4 + 2.044665X_5$$

Since the data were scaled to facilitate computation, it may be awkward to use the regression equation in its present form. In such a case, the equation can be converted to more desirable units. The first step would be to multiply each component of the equation by 100 in order to convert Y, which is currently in hundreds of bushels of corn, to Y*, which is in bushels of corn (1 bu = 35.2 liters).

$$Y^* = 44.2522 + 1.245129N - 1.043613\frac{N^2}{100} - 1.492851P -$$

$$0.357163\frac{P^2}{100} + 2.044665\frac{NP}{100}$$

Even this revised form may be somewhat awkward but can be readily transformed to the following:

$$Y^* = 44.2522 + 1.245129N - 0.01043613N^2 - 1.492851P$$
$$- 0.00357163P^2 + 0.02044665(NP)$$

And finally, the equation would be simplified through rounding to:

$$Y^* = 44.25 + 1.245N - 0.01044N^2 - 1.493P - 0.003572P^2$$
$$+ 0.02045(NP)$$

A cursory examination of the equation reveals the disturbing fact that as P increases, yield decreases. A more thorough examination of the equation indicates that the positive interaction term is measuring the positive impact on yield of increasing levels of phosphorus.

Unfortunately, b values of the wrong sign are a common problem in regression analysis. They are usually the result of multicolinearity, that is, a situation where the X_i are not truly independent of one another. In this case, N and P are independent of one another (i.e., orthogonal); however, P is not independent of the interaction term (NP). In some instances, the overall regression will be significant (as indicated by an F test) while each b will test as "not significantly different from zero." The cause of this disparity of results is multicolinearity. In other words, it is not possible to test for changes in Y given a change in a single X holding all other X's constant because as one X changes (in this case P), one or more other X's (in this case P^2 and NP) will also change. As a result, the set of X's may give a very good explanation of changes in Y while each b will test as nonsignificant.

Although it is a common practice to discard nonsignificant explanatory variables and recalculate the regression equation, this technique can result in considerable confusion when multicolinearity exists. In general, the researcher is left to theoretical considerations and his own intuition to decide what variables are important and should remain in the equation. In addition, as Brownlee (1965) (pp 452–454) has shown, omitting an important variable will bias the coefficients that are estimated. As a result, it may be better to retain a nonsignificant variable—especially if the purpose of the regression equation is to predict future values of Y. An even better approach may be to use a test of homogeneity which will be illustrated in the following section on the use of Dummy Variables.

It is left to the reader to use the data in Tables 13.12 and 13.13 to determine that $R^2 = 0.87$, to construct the appropriate ANOV table for this regression, and to calculate the Backward Solution in order to test for the significance of individual b values.

DUMMY VARIABLES

Although regression analysis is used primarily to handle quantitative data, it has the capability of handling qualitative data as well. For example, we may be interested in analyzing monthly or quarterly data for price-quantity relationships. The question then becomes one of deciding how to code months as independent variables. Alternatively, we may have observations on males and females. Again, the question arises of how to code such qualitative differences. The easiest way to handle such a problem is with dummy variables (which are sometimes referred to as zero-one variables, binary variables, or dichotomous variables).

In order to use dummy variables, the following conditions must be met:

(1) The data can be grouped into mutually exclusive classes.
(2) It is hypothesized that the effect of the qualitative difference is to shift the intercept without changing the slope. (This hypothesis can be tested.)

The general model for multiple regression can be written as:

$$Y = \mu + \beta_1 x_1 + \beta_2 x_2 + \ldots + \beta_m x_m + \epsilon$$

The model for using dummy variables can be written as:

$$Y = \mu + T_i + \beta_1 x_1 + \beta_2 x_2 + \ldots + \beta_m x_m + \epsilon$$

This is mathematically equivalent to the analysis of covariance model of Chapter 10. Although the models are identical, we may prefer the analysis of covariance format or the regression analysis with dummy variables format depending upon our point of view. In other words, if we are primarily interested in evaluating treatments and consider the covariates as a distraction, we will prefer the analysis of covariance model which focuses attention on treatment effects. On the other hand, if we are primarily interested in the relationship between Y and our various X's and consider the treatments as something that must be statistically extracted in order to clarify the relationship between Y and X, then we will prefer to use regression analysis with dummy variables.

Table 13.14 reports price and quantity of broilers by quarters for the United States during the years 1975–1977. Although we may be primarily interested in the price-quantity relationship, it is well known that the demand for broilers is affected by season of the year. Figure 13.2 shows the 3 alternative models which may be considered. Figure 13.2A indicates that one choice is a single regression line implying that there are no seasonal effects. Figure 13.1B indicates that the effect of season is a shift in the intercept which can be analyzed in a single multiple regression utilizing dummy variables. The result of such an analysis is a distinct curve for each season. Figure 13.2C indicates that the effect of season is to change the slope of the line as well as the intercept. In this last instance, the data from each quarter must be treated as a sample from a unique population. Consequently, 4 separate regressions must be performed.

TABLE 13.14. BROILER SUPPLY AND PRICES FOR THE UNITED STATES BY QUAR-
TERS, 1975–1977

Year and Quarters	Quantity Military lb	Retail Price ¢/lb
1975		
I	1833	41.2
II	2062	43.7
III	2080	50.3
IV	1992	45.1
1976		
I	2116	42.2
II	2314	41.7
III	2372	41.5
IV	2185	35.5
1977		
I	2156	40.9
II	2399	42.3
III	2424	42.4
IV	2248	37.6

1 lb = 0.45 kg.

Before the analysis can be performed, it is necessary to determine
whether price or quantity is the dependent variable. Although economic
theory states that price is the independent variable and quantity is the
dependent variable, it must be borne in mind that when dealing with
national market data, the quantity produced is normally fixed and predetermined. Consequently, quantity is the independent variable and price is
the dependent variable because within the time frame of our observations,
price is the only variable that can change.

In order to estimate the parameters for the model

$$Y = \mu + T_i + \beta(X - \bar{x}) + \epsilon$$

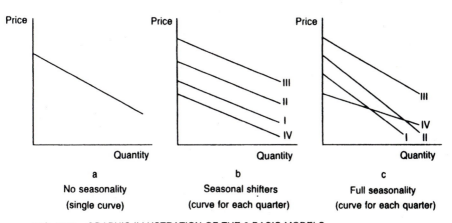

FIG. 13.2.   GRAPHIC ILLUSTRATION OF THE 3 BASIC MODELS

it is necessary to impose a restriction in order to avoid singularity in the matrix. The authors feel that one of the better methods of imposing a constraint is to state that the sum of the treatment effects is zero (i.e., $\sum_{i=1}^{k} T_i = 0$). This can be readily achieved by defining the last treatment as equal to the negative of the sum of the other treatments (i.e., $T_k = - \sum_{i=1}^{k-1} T_i$). Table 13.15 shows our 12 observations coded for a regression using dummy variables. We code a 1 when the treatment is present and zero otherwise. Since the last quarter is defined as the negative of the sum of the other quarters, the dummy variables take on the value of -1 during the fourth quarter.

TABLE 13.15.   PRICE-QUANTITY DATA OF TABLE 13.14 CODED FOR REGRESSION ANALYSIS

Y Price	Dummy Variables			Real Variable X_4 Quantity
	X_1 I	X_2 II	X_3 III	
41.2	1	0	0	1833
43.7	0	1	0	2062
50.3	0	0	1	2080
45.1	-1	-1	-1	1992
42.2	1	0	0	2116
41.7	0	1	0	2314
41.5	0	0	1	2372
35.5	-1	-1	-1	2185
40.9	1	0	0	2156
42.3	0	1	0	2399
42.4	0	0	1	2424
37.6	-1	-1	-1	2248

The regression equation derived from the data of Table 13.15 is:

$$Y = 72.91 - 2.677X_1 + 1.617X_2 + 4.260X_3 - 0.01415X_4$$

Since the effect of the fourth quarter is equal to the negative of the sum of the other 3 quarters:

$$T_4 = -(-2.677 + 1.617 + 4.260)$$

$$= -3.200$$

In other words, we could write a forecasting (or price-quantity) equation for each quarter by adding the effect of the quarter to the intercept.

I
$$P = (72.91 - 2.677) - 0.01415\,Q$$
$$P = 70.23 - 0.01415\ Q$$

II
$$P = 74.53 - 0.01415\ Q$$

III
$$P = 77.17 - 0.01415\ Q$$

IV

$$P = 69.71 - 0.01415 \, Q$$

Written in this way, it is easy to see that because of changing demand conditions during the year, the same quantity on the market will command the highest price during the third quarter and the lowest price in the fourth quarter. Of course, if we had used monthly or weekly data, the seasonal transition would be much smoother but the pattern would be the same.

Having estimated an equation with dummy variables, the question must be addressed as to whether or not the form of the model is appropriate. Table 13.16 summarizes the results of estimating the 3 alternative models illustrated in Fig. 13.2.

Using a procedure presented by Logan and Boles (1962), 3 models so constructed can be compared for homogeneity. While the sum of squared residuals (ESS) is reduced as one moves from the more restricted model A to the more general model C, there is a concurrent loss in the degrees of freedom. The Logan and Boles procedure uses this information to determine if the models have been significantly improved by allowing them to become more general. Significant improvement in models B and/or C over A indicates that there are significant seasonal variations in the demand for broilers. Similarly, B and C can be compared to determine the more satisfactory of the two.

The test is of the general form:

$$F_{(D_1, D_2)} = \frac{[S(a) - S(A)]/D_1}{S(A)/D_2}$$

where $S(a)$ = the sum of squared residuals of the more restricted model

   $S(A)$ = the sum of squared residuals of the more general model

   D_1 = the difference in the degrees of freedom for the 2 models being compared

and D_2 = the degrees of freedom for the more general model

In our case, comparing models A and B

$$F_{(3,7)} = \frac{\dfrac{(137.926 - 55.233)}{3}}{\dfrac{55.233}{7}}$$

$$= \frac{27.564}{7.890}$$

$$= 3.494$$

TABLE 13.16.   SUMMARY OF REGRESSION RESULTS FOR 3 MODELS

Model	Intercept	Equation Quarterly Effect				Q Coefficient	R^2	df	ESS
		I	II	III	IV				
A No seasonal effect	52.14					-0.00463	0.05	10	137.926
B Seasonal shifts estimated by dummy variables	72.91	-2.677	+1.617	+4.260	-3.200	-0.01415	0.62	7	55.233
C One equation for each quarter	39.97					+0.00072	0.03	1	0.894
	53.90					-0.00502	0.73	1	0.560
	102.95					-0.02540	0.95	1	2.516
	112.43					-0.03410	0.81	1	9.557

This, when compared with a tabular value for F with 3 and 7 degrees of freedom of 4.35 at the 5% level of significance, indicates that the introduction of the dummy variables did not significantly reduce the error sum of squares.

Since B did not represent a significant improvement over A, we next compare A with C. In this case

$$F_{(3,7)} = \frac{\dfrac{137.926 - (0.894 + 0.560 + 2.516 + 9.557)}{6}}{\dfrac{(0.894 + 0.560 + 2.516 + 9.557)}{4}}$$

$$= \frac{20.733}{3.382}$$

$$= 6.130$$

Since the tabular value for F with 6 and 4 degrees of freedom at the 5% level of significance is 6.16, we conclude that 4 separate equations (i.e., 1 for each quarter) do not represent a significant improvement over our hypothesis of no seasonal effect.

This simple example of the use of dummy variables also illustrates the dangers associated with small samples. In practice, we would have used monthly or weekly data for a period of 7 to 10 years. In addition, we would have used per capita quantity (instead of national quantity) to account for population growth; we would have deflated price by the consumer price index (i.e., CPI, see Chapter 17); and we would have added additional explanatory variables (e.g., per capita income, supplies of beef and pork as substitute goods, etc.) By keeping the data set small and the model extremely simple, we estimated a nonsignificant statistical relationship.

ANOV VERSUS REGRESSION ANALYSIS

Chapters 7, 8, and 9 presented the analysis of variance model in considerable detail. In Chapter 12 and in this chapter, we used the ANOV format to test the overall regression because the ANOV format provided us with a framework for partitioning the total sum of squares into a regression sum of squares and an error sum of squares. However, we could have analyzed our data directly within an analysis of variance setting. For example, Table 13.17 presents the original data from Table 13.4 in a manner suitable for constructing a typical ANOV table.

Table 13.17 reports the results of an experiment with 7 treatments (i.e., fertilizer levels) with 2 replications. Table 13.18 summarizes the results in a standard ANOV table. The tabular value for $F_{6,7}$ is 7.19 at the 1% level. Consequently, we would conclude that there are significant differences among treatments. It should be noted that the analysis of variance model

TABLE 13.17. CORN YIELD DATA OF TABLE 13.4 ORGANIZED FOR A 1-WAY ANOVA

Computation	Fertilizer Treatment							Total
	90	100	110	120	130	140	150	
	85	90	95	100	100	95	88	
	83	92	94	101	99	93	90	
$\sum_j Y_{ij}$	168	182	189	201	199	188	178	1305
$\sum_j Y^2_{ij}$	14,114	16,564	17,861	20,201	19,801	17,674	15,844	122,059
$\dfrac{\left[\sum_j Y_{ij}\right]^2}{r}$	14,112	16,562	17,860.5	20,200.5	19,800.5	17,672	15,842	122,049.5

TABLE 13.18.   ANALYSIS OF VARIANCE TABLE OF YIELD DATA

Sources of Variation	Degrees of Freedom	Sums of Squares	Mean Square	F
Treatments	6	404.857	67.476	49.724**
Error	7	9.500	1.357	—
Total	13	414.357	—	—

**Significant at the 1% level.

considers treatments as qualitative differences. As a result, 6 degrees of freedom are associated with the treatment sum of squares of Table 13.18. In contrast, by recognizing the quantitative nature of the treatments (i.e., lb N/A) and by specifying a functional relationship (i.e., $Y = a + b_1 N + b_2 N^2$), the regression analysis—summarized in Table 13.8—required only 2 degrees of freedom. In other words, if we are dealing with quantitative data and have an insight into the functional relationship, we should use regression analysis. If we have a mixture of quantitative and qualitative data, we should use regression with dummy variables or the analysis of covariance model. On the other hand, if we have only qualitative data or if the functional relationship is impossible to discern, we should use the analysis of variance model.

SUMMARY

Regression analysis is a powerful statistical tool. However, it is not a substitute for common sense or experienced judgement. Regression analysis permits the estimation of functional relationships and thus extracts a great deal of information from what can often be a confusing array of data and reduces it to a few basic statistics. In addition, if certain assumptions are valid, regression analysis permits the testing of these statistics for statistical significance. As the individual acquires experience in the use of regression analysis, he finds that it is one of the more valuable statistical tools available for both business and research.

PROBLEMS

13.1.   Show that $b^2 \Sigma x^2 = b \Sigma xy$

13.2.   Show that $\Sigma Xy = \Sigma xy$

13.3.   Show that $\Sigma xY = \Sigma xy$

13.4.   An apple sauce is being prepared from apples of varying degrees of firmness. An experiment is conducted to determine the relationship between the consistency of the final product (as measured by flow rate) on the one hand and the initial firmness (as measured by the shear value) and the amount of water added (in percentage) on the other.

The observed data follow (1 lb = 0.45 kg):

Shear Value (lb) X_1	Water Added (%) X_2	Consistency (sec) Y
300	0	45
280	5	20
310	10	10
410	2	50
405	7	35
415	12	22
520	4	65
500	9	45
480	14	30
590	6	75
600	11	60
600	16	36

(a) Calculate the regression equation

$$Y = a + b_1 X_1 + b_2 X_2$$

(b) Test each b to determine whether or not it is significantly different from zero.

(c) Summarize the regression results in an analysis of variance table which tests the following null hypothesis:

$$H_0: \beta_i = 0 \text{ for all } i$$

against the alternative

$$H_1: \beta_i \neq 0 \text{ for at least one } i.$$

REFERENCES

BROWNLEE, K.A. 1965. Statistical Theory and Methodology in Science and Engineering. John Wiley & Sons, New York.

FOOTE, R.J. 1955. A comparison of single and simultaneous equation techniques. J. Farm Econ. *37*, 975–990.

GOLDBERGER, A.S. 1964. Econometric Theory, 2nd Edition. John Wiley & Sons, New York.

JOHNSTON, J. 1972. Econometric Methods. McGraw-Hill Book Co., New York.

LOGAN, S.H. and BOLES, J.N.   1962.   Quarterly fluctuations in retail prices of meat.   J. Farm Econ. *44*, 1050–1060.

Applications of multiple regression in recent literature:

BROWN, W.F., HOLLOWAY, J.W. and BUTTS, W.T., JR.   1980.   Patterns of change in mature angus cow weight and fatness during the year.   J. Anim. Sci. *51*, 43.

HSU, W.H. and DENG, J.C.   1980.   Processing of cured mullet roe.   J. Food Sci. *45*, 97.

LAH, C.L., CHERYAN, M. and DEVOR, R.E.   1980.   A response surface methodology approach to the optimization of whipping properties of an ultra-filtered soy product.   J. Food Sci. *45*, 1720.

SCHEN, J.A., MONTGOMERY, M.W. and LIBBEY, L.M.   1980.   Subjective and objective evaluation of strawberry pomace essence.   J. Food Sci. *45*, 41.

SULLIVAN, J.F., CRAIG, J.C., JR., KONSTANCE, R.P., EGOVILLE, M.J. and ACETO, N.C.   1980.   Continuous explosion-puffing of apples.   J. Food Sci. *45*, 1550.

Suggested Further Readings

ANDERSON, R.L. and BANCROFT, T.A.   1952.   Statistical Theory in Research.   McGraw-Hill Book Co., New York.

DRAPER, N.R. and SMITH, H.   1966.   Applied Regression Analysis. John Wiley & Sons, New York.

GOLDBERGER, A.S.   1968.   Topics in Regression Analysis.   Macmillan Publishing Co., New York.

KMENTA, J.   1971.   Elements of Econometrics.   Macmillan Publishing Co., New York.

STEEL, R.G.D. and TORRIE, J.H.   1960.   Principles and Procedures of Statistics.   McGraw-Hill Book Co., New York.

WAUGH, F.V.   1964.   Demand and price analysis: Some examples from agriculture.   U.S. Dep. Agric. Tech. Bull. *1316*.

14

Correlation Analysis

Although correlation analysis is one of the most widely used statistical techniques, it is rarely used correctly. The following conditions are implied by the use of correlation analysis:

(1) The total observation is a randomly drawn sample.
(2) The sampled distribution is a multivariate normal distribution.
(3) All of the variables are dependent variables.

The first condition means that correlation is inappropriate when planned experiments are analyzed since the X's will be fixed and predetermined. The second condition takes us out of the realm of this text which considers only univariate statistics. However, it is the last requirement that creates the greatest difficulty. Correlation analysis makes no distinction between dependent and independent variables because all are considered to be dependent.

It is a common practice to design an experiment and calculate the correlation coefficients relating the dependent variable (Y) to each of the independent variables (X). But it must be borne in mind that the purpose of constructing an experiment is normally to detect and measure the underlying causal relationship between Y and the various X's which may be appropriate. As Kramer and Szczesniak (1973) (pp 135–136) point out:

In correlation analysis, the basic question is whether or not two variables move together. There is no assumption of causality. In fact, it may be that changes in the two variables may be the result of a third variable which may be unspecified. For example, there is a high correlation between the consumption of wine and the salary of teachers in the United States. Instead of concluding that the moral fiber of teachers is deteriorating, the usual explanation is that the rising levels of both are the result of rising per capita income in the United States.

In contrast, regression analysis has implicit in it the assumption of a unilateral causality. That is, changes in X result in changes in Y. However, changes in Y do not result in changes in X. This unilateral causality is basic to the mathematics underlying regression analysis. For example, a regression where the height of the son (Y) is a function of the height of the father (X) is consistent with the basic assumption of a unilateral causality. Texture measurements and the relationship between sensory and objective measurements are considerably more complex than the mathematical assumptions which underlie either correlation analysis or regression analysis. This does not mean that these statistical techniques cannot be used. Instead, it means that they must be used with great caution. In particular, tests of significance will not be valid when the underlying assumption is violated.

In general, a researcher will be interested in analyzing a cause and effect relationship. In such a circumstance, he will use regression analysis as presented in Chapters 12 and 13. In those few instances where no cause and effect relationship is implied and correlation analysis is appropriate, the following formulas may be used.

SIMPLE LINEAR CORRELATION

Figure 14.1 illustrates 4 data situations that might be observed. Figure 14.1A indicates that there seems to be a strong positive relationship between X and Y (i.e., they seem to increase or decrease together). Figure 14.1B seems to indicate no relationship between X and Y. Figure 14.1C seems to indicate a strong negative relationship between X and Y (i.e., as one increases, the other decreases). Figure 14.1D seems to indicate a strong relationship between X and Y which is nonlinear—apparently quadratic.

For any data set, the correlation coefficient (r) is calculated as follows:

$$r = \frac{\Sigma XY - \frac{(\Sigma X)(\Sigma Y)}{n}}{\left(\left[\Sigma X^2 - \frac{(\Sigma X)^2}{n}\right]\left[\Sigma Y^2 - \frac{(\Sigma Y)^2}{n}\right]\right)^{1/2}}$$

It can be shown that the mathematical limits for r are -1 and $+1$. A value of $+1$ indicates that all of the observations lie on a single line with a positive slope. A value of -1 indicates that all of the values lie on a single line with a negative slope. In Fig. 14.1, we would calculate a correlation coefficient (r) of close to $+1$ for Fig. 14.1A and close to -1 for Fig. 14.1C. Figures 14.1B and 14.1D would have correlation coefficients near zero.

An r of zero for Fig. 14.1B seems reasonable since there is no apparent relationship between X and Y. However, an r of zero for Fig. 14.1D requires a word of explanation. It is obvious from visual inspection that there is a regular pattern of behavior relating X and Y exhibited in Fig. 14.1D. Although there is a regular pattern, it is nonlinear and the value of r will be near zero. This illustrates one of the problems of using correlation. If the researcher relies only on the calculated value of r, he will miss the relationship between X and Y. Only by plotting the data on a scatter diagram similar to Fig. 14.1D will the researcher detect the nonlinear nature of the relationship.

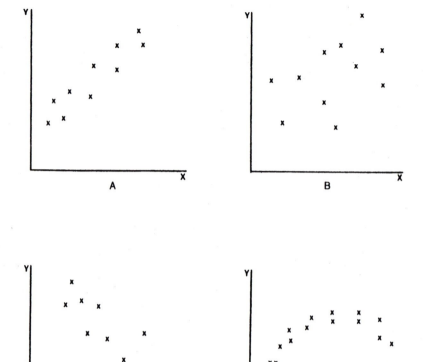

FIG. 14.1. AN ILLUSTRATION OF 4 POSSIBLE SETS OF OBSERVATIONS

As with regression analysis, r^2 indicates the percentage of the variation in Y associated with variation in X. Unlike regression analysis, r^2 (in correlation analysis) also indicates the percentage of variation in X associated with variation in Y.

PARTIAL CORRELATION

When more than two variables are involved in an analysis, it is usually of interest to calculate the partial correlation coefficients. A partial correlation coefficient indicates the relationship between two variables when all other variables are held constant.

Table 14.1 reports the results of tests performed on raw apples and applesauce made from those apples. The average panel score (Y_1) is the average reported for a sample of applesauce. The consistency (Y_2) is the average value of a physical measure from the same applesauce. The shear readings (Y_3) are the average values from raw apples sampled from the apples used to make each batch of applesauce. In a sense, these are all dependent variables and measurements of the crispness and texture of the raw apples. The data were ordered in terms of ascending panel scores.

TABLE 14.1. MEASUREMENTS ON RAW APPLES AND CANNED APPLESAUCE

Y_1 Average Panel Score $(-3$ to $+3)$	Y_2 Consistency, sec per 250 ml Flow	Y_3 Shear, lb Force
-1.9	15	245
-1.7	25	300
-1.1	23	330
-1.0	40	270
-0.7	31	390
-0.2	48	490
+0.8	65	560
+1.0	52	480
+1.5	70	600
+3.0	80	700

1 lb$_f$ = 4.448 newtons.

Table 14.2 reports the first step in our correlation analysis. Using our formula for r, the simple correlation coefficients relating each pair of variables are calculated:

$$r_{11} = \frac{\Sigma Y_1 Y_1 - \dfrac{(\Sigma Y_1)(\Sigma Y_1)}{n}}{\left(\left[\Sigma Y_1^2 - \dfrac{(\Sigma Y_1)^2}{n}\right]\left[\Sigma Y_1^2 - \dfrac{(\Sigma Y_1)^2}{n}\right]\right)^{1/2}}$$

$$= \frac{22.13 - \dfrac{(-0.3)(-0.3)}{10}}{\left(\left[22.13 - \dfrac{(-0.3)^2}{10}\right]\left[22.13 - \dfrac{(-0.3)^2}{10}\right]\right)^{1/2}}$$

$$= 1.000000$$

$$r_{12} = \frac{\Sigma Y_1 Y_2 - \dfrac{(\Sigma Y_1)(\Sigma Y_2)}{n}}{\left(\left[\Sigma Y_1^2 - \dfrac{(\Sigma Y_1)^2}{n}\right]\left[\Sigma Y_2^2 - \dfrac{(\Sigma Y_2)^2}{n}\right]\right)^{1/2}}$$

$$= \frac{281.4 - \dfrac{(-0.3)(449)}{10}}{\left(\left[22.13 - \dfrac{(-0.3)^2}{10}\right]\left[24{,}473 - \dfrac{(449)^2}{10}\right]\right)^{1/2}}$$

$$= \frac{294.87}{[(22.121)(4312.9)]^{1/2}}$$

$$= 0.954649$$

$$r_{13} = \frac{\Sigma Y_1 Y_3 - \dfrac{(\Sigma Y_1)(\Sigma Y_3)}{n}}{\left(\left[\Sigma Y_1^2 - \dfrac{(\Sigma Y_1)^2}{n}\right]\left[\Sigma Y_3^2 - \dfrac{(\Sigma Y_3)^2}{n}\right]\right)^{1/2}}$$

$$= \frac{1948.5 - \dfrac{(-0.3)(4365)}{10}}{\left[\left(22.13 - \dfrac{(-0.3)^2}{10}\right)\left(2{,}118{,}025 - \dfrac{(4365)^2}{10}\right)\right]^{1/2}}$$

$$= \frac{2079.45}{[(22.121)(212{,}702.5)]^{1/2}}$$

$$= 0.958649$$

TABLE 14.2.   SIMPLE CORRELATION COEFFICIENTS RELATING PANEL SCORE (Y_1), CONSISTENCY (Y_2), AND SHEAR (Y_3)

	Y_1	Y_2	Y_3
Y_1	1.000000	0.954649	0.958649
Y_2	0.954649	1.000000	0.942501
Y_3	0.958649	0.942501	1.000000

The remaining values for Table 14.2 are calculated in a similar manner. It should be noted that the matrix of correlation coefficients is symmetric (i.e., $r_{ij} = r_{ji}$). In addition, a variable will always correlate perfectly with itself (i.e., $r_{ii} = 1.0$).

It is clear from the results in Table 14.2 that these variables are highly correlated with one another. Partial correlation addresses the question of the relationship between two variables holding all other variables constant. In order to calculate partial correlation coefficients, it is necessary to invert the matrix of correlation coefficients shown in Table 14.2. This can be accomplished using the Abbreviated Doolittle Method illustrated in Chapter 13. However, if it is known that a matrix is to be inverted (i.e., the forward plus the backward solution of the Abbreviated Doolittle), it may be simpler to invert the matrix through the method known as "row reduction."

Table 14.3 shows the inversion of the matrix shown in Table 14.2. In essence, this procedure takes two matrices (A and B) where the A matrix is the matrix to be inverted and the B matrix is an identity matrix (i.e., a matrix with ones on the main diagonal and zeros elsewhere). The A matrix is subjected to a series of arithmetic steps until it is transformed into an identity matrix. If these same steps are imposed on the B matrix, the final B matrix will be the inverse of the original A matrix.

Table 14.3 was constructed as follows:

(1) The original A matrix was copied from Table 14.2. The B (identity) matrix is defined to the right of the A matrix.

(2) The second step begins by ensuring that the diagonal element in the A matrix is 1.0. This can be accomplished by dividing each element in the row by the diagonal element of the A matrix. Since at our first iteration this is already true, the first line is left unchanged. Our next step is to ensure that all other elements in the first column of the A matrix are 0.0. This is achieved in the following manner:

(a) We note that $a_{21} = 0.954649$. If we multiply that constant times each element in the first row and subtract the product from the second row, we will zero out the cell a_{21} and adjust all other items in the second row accordingly.

$$a_{21} = 0.954649 - (1 \times 0.954649)$$
$$= 0.0$$
$$a_{22} = 1.0 - (0.954649 \times 0.954649)$$
$$= 0.088645$$
$$a_{23} = 0.942501 - (0.958649 \times 0.954649)$$
$$= 0.027328$$

$$b_{21} = 0.0 - (1.0 \times 0.954649)$$
$$= -0.954649$$
$$b_{22} = 1.0 - (0.0 \times 0.954649)$$
$$= 1.0$$
$$b_{23} = 0.0 - (0.0 \times 0.954649$$
$$= 0.0$$

TABLE 14.3.   MATRIX INVERSION BY MEANS OF ROW REDUCTION

Instruction	A Matrix			Identity Matrix (B)		
1	1.	0.954649	0.958649	1.	0	0
	0.954649	1.	0.942501	0	1.	0
	0.958649	0.942501	1.	0	0	1.
2	1.	0.954649	0.958649	1.	0	0
	0.0	0.088645	0.027328	-0.954649	1.0	0.0
	0.0	0.027328	0.080992	-0.958649	0.0	1.0
3	1.	0.0	0.664344	11.280949	-10.76935	0
	0.0	1.0	0.308286	-10.76935	11.280952	0.0
	0.0	0.0	0.072567	-0.664344	-0.308286	1.0
4	1.0	0.0	0.0	17.362955	-7.947021	-9.154905
	0.0	1.0	0.0	-7.947021	12.590642	-4.248295
	0.0	0.0	1.0	-9.154905	-4.248295	13.780368

(b) We note that $a_{31} = 0.958649$. If we multiply that constant times each element in the first row and subtract the product from the third row, we will zero out the cell a_{31} and adjust all other items in the third row accordingly.

$$a_{31} = 0.958649 - (1.0 \times 0.958649)$$
$$= 0.0$$
$$a_{32} = 0.942501 - (0.954649 \times 0.958649)$$
$$= 0.027328$$
$$a_{33} = 1.0 - (0.958649 \times 0.958649)$$
$$= 0.080992$$

$$b_{21} = 0 - (1.0 \times 0.958649)$$
$$= -0.958649$$
$$b_{22} = 0.0 - (0.0 \times 0.958649)$$
$$= 0.0$$
$$b_{23} = 1.0 - (0.0 \times 0.958649)$$
$$= 1.0$$

We have now converted the A matrix into a semblance of an identity matrix by achieving a first column with a 1 on the main diagonal and zeros for the remaining elements.

(3) The third step proceeds by converting the element on the main diagonal of the second row (i.e., a_{22}) into a 1. This is accomplished by dividing each element of the second row of the second instruction by the constant 0.088645.

$$0.0 = 0.0 \div 0.088645$$
$$1.0 = 0.088645 \div 0.088645$$
$$0.308286 = 0.027328 \div 0.088645$$
$$-10.76935 = -0.954649 \div 0.088645$$
$$11.280952 = 1.0 \div 0.088645$$
$$0.0 = 0.0 \div 0.088645$$

We continue as before by zeroing out the remaining elements of the second column.

(a) We note that $a_{12} = 0.954649$. If we multiply that constant times each element in the second row and subtract the product from the first row, we will zero out the cell a_{12} and adjust all other items in the first row accordingly.

$$a_{11} = 1.0 - (0.0 \times 0.954649)$$
$$= 1.0$$
$$a_{12} = 0.954649 - (1.0 \times 0.954649)$$
$$= 0.0$$
$$a_{13} = 0.958649 - (0.308286 \times 0.954649)$$
$$= 0.664344$$

$$b_{11} = 1.0 - (-10.76935 \times 0.954649)$$
$$= 11.280949$$

$$b_{21} = 0.0 - (11.280952 \times 0.954649)$$
$$= -10.76935$$
$$b_{31} = 0.0 - (0.0 \times 0.954649)$$
$$= 0$$

(b) Using the same procedure, the third row is adjusted.

$$a_{31} = 0.0 - (0.0 \times 0.027328)$$
$$= 0.0$$
$$a_{32} = 0.027328 - (1.0 \times 0.027328)$$
$$= 0.0$$
$$a_{33} = 0.080992 - (0.308286 \times 0.027328)$$
$$= 0.072567$$

$$b_{31} = -0.958649 - (-10.76935 \times 0.027328)$$
$$= - 0.664344$$
$$b_{32} = 0.0 - (11.280952 \times 0.027328)$$
$$= -0.308286$$
$$b_{33} = 1.0 - (0.0 \times 0.027328)$$
$$= 1.0$$

(4) The fourth step begins by converting the element on the main diagonal to a 1. The procedure continues using the same technique illustrated in steps (2) and (3).

When the A matrix has been transformed into an identity matrix, the B matrix is the inverse of the original matrix. The inverse, which is shown in the lower right of Table 14.3, is reported in Table 14.4.

TABLE 14.4.   INVERSE OF THE MATRIX SHOWN IN TABLE 14.2

17.362955	−7.947021	−9.154905
−7.947021	12.590642	−4.248295
−9.154905	−4.248295	13.780368

Given the inverse of the correlation matrix, any partial correlation coefficient can be calculated with the following formula:

$$r_{ij\cdot1,2,3,\cdots,i\,-\,1,i\,+\,1,\cdots,j\,-\,1,j\,+\,1,\cdots,k} = \frac{-b_{ij}}{[b_{ii}b_{jj}]^{1/2}}$$

For our data:
The partial correlation between Y_1 and Y_2 given Y_3 is:

$$r_{12\cdot3} = \frac{-b_{12}}{(b_{11}b_{22})^{1/2}}$$

$$= \frac{-(-7.947021)}{(17.362955 \times 12.590642)^{1/2}}$$

$$= 0.537$$

The partial correlation coefficient between Y_1 and Y_3 given Y_2 is:

$$r_{13\cdot2} = \frac{-(-9.154905)}{(17.362955 \times 13.780368)^{1/2}}$$

$$= 0.592$$

The partial correlation coefficient between Y_2 and Y_3 given Y_1 is:

$$r_{23\cdot1} = \frac{-(-4.248295)}{(12.590642 \times 13.780368)^{1/2}}$$

$$= 0.323$$

Each of the simple correlations was very high. The partial correlation coefficients are much smaller. Since the 3 variables move together, holding 1 constant means that the movements of the other 2 are also restricted. In addition, with all of the values in the A matrix nearly equal to 1.0, there is the potential problem of singularity in the matrix. In such a case, the estimates of the partial correlation coefficients may be quite unstable. In other words, repeating the experiment and recalculating the answers could yield very different estimated partial correlation coefficients even though the same multivariate normal population is being sampled.

SUMMARY

Correlation is a useful statistical technique. When used with caution and judgment, it is a valuable tool for beginning an analysis. Because of the similarity of calculations and notation, there is considerable confusion between correlation and regression. It should always be borne in mind that regression analysis implies a causal relationship; correlation analysis implies that there is *no* causal relationship.

PROBLEMS

14.1. Plot the scatter diagrams for $Y_1{:}Y_2$, $Y_1{:}Y_3$, and $Y_2{:}Y_3$ for the data in Table 14.1.

14.2. Using the matrix in Table 14.4
    (a) Calculate $r_{21 \cdot 3}$
    (b) Does $r_{21 \cdot 3} = r_{12 \cdot 3}$?
    (c) Calculate $r_{31 \cdot 2}$
    (d) Does $r_{31 \cdot 2} = r_{13 \cdot 2}$?

REFERENCES

DANIEL, W.W. and TERRELL, J.C. 1975. Business Statistics: Basic Concepts and Methodology. Houghton Mifflin Co., Boston.

KRAMER, A. and SZCZESNIAK, A.S. 1973. Texture Measurements of Foods. D. Riedel Publishing Co., Dordrecht, Holland.

STEEL, R.G.D. and TORRIE, J.H. 1960. Principles and Procedures of Statistics. McGraw-Hill Book Co., New York.

Applications of correlation analysis in recent literature:

ADAMS, N.J., EDWARDS, R.L., SMITH, G.C., RIGGS, J.K. and CARPENTER Z.L. 1980. Performance and carcass traits of progeny of imported and domestic hereford bulls. J. Anim. Sci. 51, 270.

CHUNG, O.K., POMERANZ, Y., JACOBS, R.M. and HOWARD, B.G. 1980. Lipid extraction conditions to differentiate among hard red winter wheats that vary in breadmaking. J. Food Sci. 45, 1168.

FOX, J.D., WOLFRAM, S.A., KEMP, J.D. and LANGLOIS, B.E. 1980. Physical, chemical, sensory, and microbiological properties and shelf life of pse and normal pork chops. J. Food Sci. 45, 786.

HAYWARD, L.H., HUNT, M.C., KASTNER, C.L. and KROPF, D.H. 1980. Blade tenderization effects on beef longissimus sensory and Instron textural measurements. J. Food Sci. 45, 925.

McMILLIN, K.W., SEBRANEK, J.G., RUST, R.E. and TOPEL, D.G. 1980. Chemical and physical characteristics of frankfurters prepared with mechanically processed pork product. J. Food Sci. 45, 1455.

15

Nonparametric Tests

The preceding chapters have dealt with procedures that are sometimes referred to as classical statistics. In each case an underlying distribution was known (or assumed) and the requisite population parameters were known or estimated. There is a growing body of literature commonly referred to as nonparametric statistics. Some authors prefer the term distribution-free statistics. Currently, the two terms are often used interchangeably. The important aspect is that some of the restrictive assumptions are relaxed. In actuality, some of these tests deal with parameters without specifying an underlying distribution. Consequently, they should be classed as distribution-free rather than nonparametric. However, such fine distinctions of definition are not important in practice.

THE RUNS TEST

The makers of Teresa's Tempting Tarts observed that, during a single 30 min selling period, apple (A) and cherry (C) tarts were sold in the following order:

AACCAAACCACCCA

The question under consideration is whether this pattern is random or nonrandom. The occurrence of nonrandomness would result in a search for the cause of nonrandomness and hopefully improved managerial decisions.

Certainly if the 14 sales had occurred in either of the following patterns:

ACACACACACACAC

or

AAAAAAACCCCCCC

management would conclude that a regular (i.e., nonrandom) pattern of behavior existed.

The runs test provides a means of detecting the existence of nonrandom behavior in cases that are not obvious.

The observed data had 7 runs.

$$\underline{AA} \quad \underline{CC} \quad \underline{AAA} \quad \underline{CC} \quad \underline{A} \quad \underline{CCC} \quad \underline{A}$$

The expected number of runs for a series is:

$$E(R) \;=\; 1 + \frac{2n_1 n_2}{n_1 + n_2}$$

where n_1 = the number of observations of the first type (in this case apple tarts)

n_2 = the number of observations of the second type (in this case, cherry tarts)

The variance of R is:

$$Var(R) \;=\; \frac{2n_1 n_2 (2n_1 n_2 - n_1 - n_2)}{(n_1 + n_2)^2 (n_1 + n_2 - 1)}$$

Although tables exist giving the critical values for several levels of significance (Daniel and Terrell 1975), in practice we can use an approximation.

The standard normal deviate is defined as:

$$Z \;=\; \frac{Y - \mu}{\sigma}$$

In this case

$$Z \;=\; \frac{R - E(R)}{[Var(R)]^{1/2}}$$

or

$$Z \;=\; \frac{R - \left[1 + \dfrac{2n_1 n_2}{n_1 + n_2} \right]}{\left[\dfrac{2n_1 n_2 (2n_1 n_2 - n_1 - n_2)}{(n_1 + n_2)^2 (n_1 + n_2 - 1)} \right]^{1/2}}$$

For our problem

$$
\begin{aligned}
R &= \text{7 (the observed number of runs)} \\
n_1 &= \text{7 (the number of apple tarts)} \\
n_2 &= \text{7 (the number of cherry tarts)}
\end{aligned}
$$

$$Z = \frac{7 - \left[1 + \dfrac{2 \cdot 7 \cdot 7}{7 + 7}\right]}{\left[\dfrac{2 \cdot 7 \cdot 7 (2 \cdot 7 \cdot 7 - 7 - 7)}{(7+7)^2 (7+7-1)}\right]^{1/2}}$$

$$Z = \frac{7 - \left[1 + \dfrac{98}{14}\right]}{\left[\dfrac{98 \ (84)}{96 \ (13)}\right]^{1/2}}$$

$$= \frac{7 - 8}{\left[\dfrac{8232}{2548}\right]^{1/2}}$$

$$= \frac{-1}{1.79743}$$

$$= -0.556$$

As can be seen from Appendix Table B.1, a standard normal deviate of 0.56 is not a rare occurrence. If Z had been greater than 1.96 (less than -1.96), we would have rejected the null hypothesis of randomness and concluded that the runs observed were generated in a nonrandom manner.

For the pattern
ACACACACACACAC

$$R = 14$$
$$n_1 = 7$$
$$n_2 = 7$$

$$Z = \frac{14 - 8}{1.79743}$$

$$= -3.338$$

For the pattern
AAAAAAACCCCCCC

$$R = 2$$
$$n_1 = 7$$
$$n_2 = 7$$

$$Z = \frac{2 - 8}{1.79743}$$

$$= -3.338$$

In either of these cases, common sense and our test statistic indicate that we should reject the null hypothesis of the random generation of runs.

The runs test is applicable to any ordered series. Ordinarily, the series will be ordered by time. The test can be applied to qualitative data (e.g., runs of males and females in a ticket line) or quantitative data which are converted to qualitative values (e.g., net weights of packages on a processing line converted to those that are above or below a specific value).

TESTS BASED ON RANKS

Occasionally data can be ranked (i.e., given ordinal values) even though cardinal values cannot be established. For example, consumers may be able to state that they prefer product A to product B (i.e., rank them) even if they are unable to state that product A is twice as good as product B. Tests based upon ranks offer the opportunity to analyze data that are only available in rank form and hence unsuitable for classical statistics. In addition, data which are originally cardinal values, where the recorded values may be suspect but their relative differences valid or where the underlying distribution is unknown, may be converted to ranks and analyzed with nonparametric tests based on ranks.

Spearman's Coefficient of Rank Correlation

One of the earliest tests based on ranks was that proposed by Spearman (1904). The calculation of an ordinary correlation coefficient is based on the assumption that the sample is drawn from a population that is bivariate normal. A correlation of ranks does not require such a restrictive assumption.

One of the uses of Spearman's coefficient of rank correlation is to evaluate taste testers. For example, if food samples are prepared with increasing percentages of sugar (i.e., the true ranks are known), we can test whether or not the human agent can correctly rank the samples.

Spearman's coefficient of rank correlation is calculated as:

$$r_s = 1 - \frac{6 \sum_i d_i^2}{n(n^2 - 1)}$$

where d_i = difference in ranks for the ith pair
     n = the number of pairs of ranked observations

When n is relatively large (e.g., 20+), the following test statistic may be used:

$$t = r_s \left[\frac{n - 2}{1 - r_s^2} \right]^{1/2}$$

which is distributed as Student's t with n − 2 degrees of freedom when $\rho = 0$ (i.e., the true correlation is zero).

The use of this statistic is illustrated in Table 15.1. In this case the true rankings are compared with the rankings developed by a taster. In the event of ties, the average rank is given. The procedure continues with the calculation of the d_i's and the d_i^2's. Of course, $\Sigma d_i = 0$. Using our formula:

$$r_s = 1 - \frac{6(4.50)}{12(144 - 1)}$$

$$= 1 - \frac{27}{1716}$$

$$= 0.984$$

TABLE 15.1. CALCULATION OF SPEARMAN'S COEFFICIENT OF RANK CORRELATION

True Ranking	Taster's Subjective Ranking	d_i (Rank Difference)	d_i^2
12	12	0.0	0.00
11	10	1.0	1.00
10	11	−1.0	1.00
9	9	0.0	0.00
8	8	0.0	0.00
7	6.5	0.5	0.25
6	6.5	−0.5	0.25
5	5	0.0	0.00
4	4	0.0	0.00
3	3	0.0	0.00
2	1	1.0	1.00
1	2	−1.0	1.00
Sum		0.0	4.50

This appears to be a significant relationship. Although our sample size is somewhat smaller than is desirable, the calculation of our test statistic supports our intuitive feeling that these results are significant.

$$t = 0.984 \left[\frac{12 - 2}{1 - 0.968} \right]^{1/2}$$

$$= 17.394$$

The limits on Spearman's coefficient of rank correlation are the same as those for the ordinary coefficient of correlation (i.e., $-1.0 \leq r_s \leq 1.0$). The special formula for r_s arises because, if the data are ranked, the shortcut formula can be used. However, using the ordinary formula for r (given in Chapter 14) will yield the same numerical values if we work directly with the ranks as our values for X and Y.

Kramer's Rank Sum Test

In 1956, Kramer proposed a test based on rank sums as an alternative to the analysis of variance model when normality could not be assumed. This work was subsequently revised and expanded tables prepared (Kramer *et al.* 1974).

This test is based on the following formula:

$$P[s = n] = \left[\frac{n!}{r!(n - r)!} \right] t^{-r}$$

where s = the sum of the ranks for a single treatment
         t = the number of treatments
         r = the number of replications

and n is varied in order to calculate the probability of each tail of the distribution

Appendix Tables I.1 and J.1 report the critical values for 5 and 1% levels of significance for $2 \leqslant r \leqslant 19$ and $2 \leqslant t \leqslant 20$.

In spite of the rather formidable calculations required to construct the tables, the procedure is quick and easy to use and is readily understood once it is illustrated.

Five formulations for a new gelatin dessert were developed. Eight panelists tasted the samples and scored each sample on a hedonic scale (1 = undesirable; 9 = highly desirable). These scores are reported in Table 15.2. The original scores were converted to ranks and summed in Table 15.3.

The null hypothesis is that there are no significant differences among treatments (in this case, formulations).

TABLE 15.2. PANELISTS' SCORES FOR 5 FORMULATIONS OF A GELATIN DESSERT

Panelist	Formulation				
	1	2	3	4	5
1	9	7	2	6	7
2	8	9	4	6	7
3	8	8	3	7	6
4	9	8	5	7	6
5	9	9	6	7	8
6	7	6	1	4	3
7	7	6	3	4	5
8	9	8	5	7	7

TABLE 15.3.   PANELISTS' SCORES OF TABLE 15.2 CONVERTED TO RANKS

Panelist	Formulation				
	1	2	3	4	5
1	1	2.5	5	4	2.5
2	2	1	5	3	4
3	1.5	1.5	5	3	4
4	1	2	5	3	4
5	1.5	1.5	5	4	3
6	1	2	5	3	4
7	1	2	5	4	3
8	1	2	5	3.5	3.5
Rank sum	10	14.5	40	27.5	28

The tables developed by Kramer (1974) list critical values for two alternative hypotheses:

(1)   There is a significant difference (i.e., at least one treatment differs).
(2)   A predetermined treatment differs significantly from the others.

For each treatment-replication combination, 4 values are reported. The upper pair report the largest and smallest insignificant rank sum for the first alternative hypothesis. The lower pair report the largest and smallest insignificant rank sum for the second alternative hypothesis.

In our case, we had 5 treatments (formulations) and 8 replications (panelists).

Using Appendix Table I.1, the critical values listed at the 5% level of significance for 5 treatments and 8 replications are:

$$15-33$$
$$17-31$$

If our alternative hypothesis is that at least 1 treatment differs from the others, we must observe a rank sum smaller than 15 or greater than 33. In Table 15.3, we observe 2 rank sums smaller than 15, either of which would cause us to reject the null hypothesis. In addition, we observe a rank sum greater than 33. This, by itself, would have resulted in our rejecting the null hypothesis. ·

However, suppose that formulation #5 was our own gelatin dessert, while the other formulations were those of our competitors. In such a case the null and alternative hypotheses would be:

H_o: Our gelatin is not significantly different from our competitors.
H_i: Our gelatin is significantly different from our competitors.

Since we are dealing with a predetermined treatment, we use the lower pair of values. The rank sum for our product (#5) is 28. This lies within the values 17–31. Consequently, we would accept the null hypothesis that our product is not significantly superior (inferior) to that of our competitors.

The previous example could have used the analysis of variance model to analyze the data, although there were apparently some problems associated with calibration of the panelists. The following example of a before and after comparison is not readily amenable to evaluation with standard parametric procedures but can be readily analyzed using Kramer's rank sum test.

Economic theory indicates that the price of a product should be highest in a central market and decline as one moves away from the market, the declining price reflecting the cost of transporting from more distant points to the central market. However, institutional barriers can prevent markets from behaving in this rational manner.

Lewis and Strand (1978) examined market prices in Maryland for oysters before and after county residential requirements were declared unconstitutional. These results are summarized in Table 15.4. Since Anne Arundel is closer to the Baltimore market, it should consistently receive the highest price for oysters, *Ceteris paribus*. However, as Table 15.4 shows, Anne Arundel received the highest price (i.e., ranked first) only 3 out of 7 years prior to the 1971 court decision. Using Appendix Table I.1 for 3 treatments, 7 replications, and a predetermined treatment (i.e., Anne Arundel should have a significantly lower rank sum) we observe the following critical values:

$$10-18$$

Since the rank sum for Anne Arundel (i.e., 14) is not less than 10, we conclude that prices in the Anne Arundel market were not significantly higher than those in the other markets at the 5% level.

TABLE 15.4.   RANKS OF MARKET PRICES FOR MARYLAND OYSTERS BEFORE AND AFTER COUNTY RESIDENT REQUIREMENTS WERE DECLARED UNCONSTITUTIONAL

Year	Before Court Case Market			Year	After Court Case Market		
	Anne Arundel	Talbot	Queen Anne		Anne Arundel	Talbot	Queen Anne
1964	1	2	3	1972	2	3	1
1965	1	3	2	1973	1	3	2
1966	3	1	2	1974	1	2	3
1967	1	3	2	1975	1	2	3
1968	3	2	1	1976	1	2	3
1969	2	3	1				
1970	3	2	1				
Rank sums	14	16	12		6	12	12

The change in market ranks after the court case appears to be very striking. Using Appendix Table I.1 for 3 treatments, 5 replications, and a predetermined treatment (i.e., Anne Arundel), we observe the following critical values:

$$7 - 13$$

Since the rank sum for Anne Arundel (i.e., 6) is smaller than the smallest nonsignificant rank sum (i.e., 7), we conclude that prices at the Anne Arundel market were significantly higher than those in the other markets after the court rendered its decision.

Kruskal-Wallis Rank Sum Test

An alternative to the Kramer rank sum test is that proposed by Kruskal and Wallis (1952). This test is especially appropriate when the replications per treatment are unequal.

The Kruskal-Wallis test statistic is calculated as:

$$K = \frac{12}{N(N + 1)} \left(\sum_{j=1}^{m} \frac{T_j^2}{n_j} \right) - 3(N + 1)$$

where K is distributed approximately as chi-square with $m - 1$ degrees of freedom

N = the total number of observations
T_j = the rank sum for the jth treatment
n_j = the number of replicates for the jth treatment
m = the number of treatments

The null and alternative hypotheses are:

H_o: There are no differences among treatments.
H_1: At least one treatment differs from the others.

The procedure begins by pooling all of the observations and ranking from least to greatest. Ties are given the average rank.

Three vendors were supplying glass jars. Table 15.5 reports the defects per 10,000 jars for 17 shipments. These 17 shipments were ordered from least to greatest and ranked. The ranks are shown in Table 15.5.

Our test statistic is:

$$K = \frac{12}{17(17 + 1)} [480.571 + 370.562 + 532.042] - 3(17 + 1)$$

$$K = \frac{12}{306} [1383.175] - 3(18)$$

$$K = 54.242 - 54$$

$$K = 0.242$$

At the 5% level of significance, chi-square with 2 degrees of freedom has a critical value of 5.991. Since the calculated value of K is less than 5.991, we accept the null hypothesis and conclude that there are no significant differences among the 3 vendors.

TABLE 15.5. THE USE OF THE KRUSKAL-WALLIS RANK SUM TEST TO DETECT DIFFERENCES AMONG VENDORS

Item	Vendor A No. of Defects	Rank	Vendor B No. of Defects	Rank	Vendor C No. of Defects	Rank
	0	1	1	2.5	3	5.5
	10	15	9	14	5	8.5
	5	8.5	6	11	4	7
	1	2.5	6	11	12	17
	11	16			7	13
	2	4			3	5.5
	6	11				
T_j		58		38.5		56.5
T^2		3364		1482.25		3192.25
n_j		7		4		6
$\dfrac{T_j^2}{n_j}$		480.571		370.562		532.042

Summary

There are a number of rank tests available. This section has presented three of them. Spearman's coefficient of rank correlation is especially useful when dealing with paired observations. Kramer's rank sum test is a quick and easy to use procedure when all treatments have equal replication. The Kruskal-Wallis rank sum test can be used in situations that have unequal as well as equal replication.

CHEBYSHEV'S INEQUALITY

A useful distribution-free procedure is that provided by Chebyshev's Inequality.

$$P(|Y - \mu| > K\sigma) \leq \frac{1}{K^2}$$

This inequality is valid for any distribution with a finite variance. It states that the probability that any random variable will be more than K standard deviations from the mean will be less than or equal to $1/K^2$.

If the distribution is known, we should use that knowledge. For example, in the normal distribution, the probability that Y will be more than 2σ from the mean is 0.046 (see Appendix Table B.1). However, Chebyshev's Inequality is able only to state that the probability is less than or equal to 0.25. That is

$$P(|Y - \mu| > 2\sigma) \leq \frac{1}{2^2}$$

Consequently, knowing the underlying distribution permits us to make a much stronger statement concerning the probability of an event.

When the underlying distribution is unknown, Chebyshev's Inequality permits us to make tests of hypotheses as long as we possess an estimate of the mean and variance. Since Chebyshev's Inequality is a general statement about any random variable Y, we may substitute $\bar{Y}$ as long as we use the appropriate variance (i.e., σ^2/n).

A common use of Chebyshev's Inequality is to determine sample size when the underlying distribution is unknown but a specified degree of accuracy is required.

A potato shipper is distributing 50 lb (22.7 kg) cartons of potatoes. He wishes to estimate the mean weight within 0.2σ with a probability of 0.95. Using Chebyshev's Inequality:

$$P(|\bar{Y} - \mu| > K\sigma_{\bar{x}}) \leq \frac{1}{K^2}$$

$$P\left(|\bar{Y} - \mu| > \frac{\sigma}{5}\right) \leq 0.05$$

where $K\sigma_{\bar{x}} = \dfrac{\sigma}{5}$

$K\dfrac{\sigma}{\sqrt{n}} = \dfrac{\sigma}{5}$

$K = \dfrac{\sqrt{n}}{5}$

Since

$$0.05 = \frac{1}{K^2}$$

$$0.05 = \frac{25}{n}$$

$$n = 500$$

Using Chebyshev's Inequality and without specifying σ, we determine that a sample of 500 observations is required in order to achieve the desired accuracy. However, if we have a preliminary estimate of σ and are willing to assume normality, we could achieve the same accuracy with a smaller sample. In this sense, Chebyshev's Inequality provides a conservative procedure.

PROBLEMS

15.1. The makers of Anthony's Animated Anchovies pack a can with a specified net weight of 7 oz (198.4 g). The line is set to pack with a mean weight of 7.1 oz (201.3 g). A series of 20 cans were removed from the line and net weights were determined. Observations were divided into those observations above the mean (A) and those below the mean (B). The following results were obtained for the series:

ABAAABAAABBBAAAAABBB

Should the firm accept the null hypothesis that the runs were generated randomly?

15.2. The following 14 pairs of observations were recorded:

X	Y
31	52
16	46
13	37
38	57
47	60
55	58
17	36
8	27
10	20
5	5
7	14
58	62
72	61
78	67

(a)   Plot the data.
(b)   Calculate the coefficient of correlation (r).
(c)   Convert the data to ranks and calculate Spearman's coefficient of rank correlation (r_s).

REFERENCES

DANIEL, W.W. and TERRELL, J.C.   1975.   Business Statistics: Basic Concepts and Methodology.   Houghton Mifflin Co., Boston.

KRAMER, A.   1956.   A quick rank test for significance of differences in multiple comparisons.   Food Technol. *14*, 391.

KRAMER, A. *et al.*   1974.   A non-parametric ranking method for the statistical evaluation of sensory data.   Chem. Senses Flavor *1*, 121−133.

KRUSKAL, W.H. and WALLIS, W.A.   1952.   Use of ranks in one-criterion variance analysis.   J. Am. Stat. Assoc. *47*, 583−612.

LEWIS, T.B. and STRAND, I.E.   1978.   Douglas *v.* Seacoast Products, Inc.: The legal and economic consequences for the Maryland oystery.   Md. Law Rev. *38* (1) 1−36.

SPEARMAN, C.   1904.   The proof and measurement of association between two things.   Am. J. Psychol. *15*, 72−101.

Applications of nonparametric analysis in recent literature:

DUNN, N.A. and HEATH, J.L.   1979.   Effect of microwave energy on poultry tenderness.   J. Food Sci. *44*, 339.

GILL, T.A., KEITH, R.A. and LALL, B.S.   1979.   Textural deterioration of red hake and haddock muscle in frozen storage as related to chemical parameters and changes in myofibrillar proteins.   J. Food Sci. *44*, 661.

IMOTO, E.M., LEE, C. and RHA, C.   1979.   Effect of compression ratio on the mechanical properties of cheese.   J. Food Sci. *44*, 343.

Suggested Further Readings

GIBBONS, J.D.   1971.   Nonparametric Statistical Inference.   McGraw-Hill Book Co., New York.

HOLLANDER, M. and WOLFE, D.A.   1973.   Nonparametric Statistical Methods.   John Wiley & Sons, New York.

LEHMANN, E.L.   1975.   Nonparametrics: Statistical Methods Based on Rank.   Holden-Day, San Francisco.

Part IV

Statistics in Business

16

Evolutionary Operations

Although statistical methods have been widely embraced by researchers, the use of statistical techniques as an aid in day-to-day production runs has been more limited. Among the barriers to widespread use are the apparent complexity of the calculations used in statistics, the restrictions imposed by the need for randomization, and the belief that only a statistician can understand the results. Box (1957) presented a technique known as evolutionary operations, or EVOP, specifically to overcome these difficulties. Evolutionary operations can be used on the production line with existing personnel. The calculations can be performed on the spot with the use of a simple worksheet, and the results provide a continuing flow of useful information to the production supervisor.

The underlying logic of EVOP is that a response surface exists similar to that explored in the traditional ANOV or regression models. That is, there is a variable (e.g., yield, color, texture) which is to be optimized that is a function of our process variables (e.g., time, temperature, line speed). By utilizing the structure of a 2×2 factorial experiment and stripping away the statistical elegance of a regular ANOV model, EVOP provides a simplified procedure, readily used by plant personnel, to systematically search the hypothesized response surface.

THE EVOP THEORY

It is seldom efficient to run an industrial process to produce product alone, but rather it should be run so as to generate product plus information on how to improve the product and reduce cost. This is the underlying philosophy of the EVOP theory, that is, to continuously probe for the opportunity for change which would result in an improvement. This is directly opposite and complementary to the control-chart approach, discussed in Chapter 19, which maintains a given process within certain limits.

Searching for an improved product or process by EVOP is accomplished by slightly displacing each of the variables under study from its normal standard operating level, first higher then lower according to a fixed pattern. The variables under study are changed each time so slightly that even if the test did not produce a better product or process, the resultant product would not be unsalable. Thus the risk of making large quantities of nonconforming product is minimized.

One of the major advantages of EVOP is that it can be performed on the production line with a minimum of interruption. Other advantages are that it does not require a thorough knowledge of mathematics and that the results are available immediately to the people who can use and analyze them, the line personnel.

Similarly as with control charts, EVOP can be applied readily to a one- or two-variable situation although the statistical procedure on which it is based, mapping of response surfaces, can be applied to situations of any degree of complexity involving more than two variables (Ferrell 1956; Box 1957; Duncan 1959).

It should be reemphasized at this point that any such applications, whether they are control charts or EVOP slopes of steepest ascent, when used for quality control purposes, should be applied "on the line" rather than in a distant laboratory research establishment. It is therefore recommended that EVOP procedures be limited to studies of one or two variables only rather than extended into broad studies of all possible variables.

THE EVOP PROCEDURE

As stated, the heart of the EVOP procedure is to probe constantly in all directions in order to find another set of operating conditions that might result in an improvement. Such changes in operation may be dictated by changes in quality of material, condition of equipment, personnel, or some unassignable cause. In any plant, there are several variables that might need investigation such as time, temperature, concentration, pH, etc.

In such studies, we may conveniently consider two factors simultaneously, such as the effect of time and temperature on yield. We always begin with the current process. After making the necessary observations, we change both factors slightly in one direction, for example, at a higher level than the current process. Then leaving factor A at the high level, adjust factor B so that it is lower than the current process. We then reduce factor A to a low level also. Finally, as the fifth treatment, we maintain the low level for A, but raise B to a high level. These treatments are shown graphically in Fig. 16.1.

It should be brought out that this procedure is not one which should be entered into without initially considering with utmost care just where the minimum point of quality is located in order not to reduce significantly the marketability of the product. It is important that the most knowledgeable people be brought into the planning stage to help determine this minimal level for each variable. It is recommended that a permanent EVOP committee be established at the plant, not only to assist in determining the level of variables, but to periodically review the results.

Thus, for example, if the time and temperature of processing are factors under consideration, a continuous study can be undertaken with the pres-

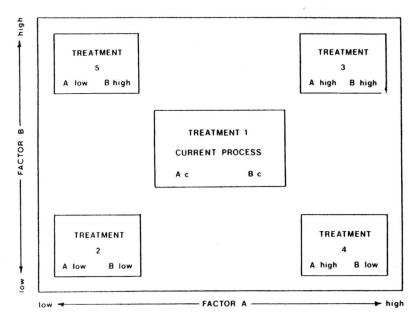

FIG. 16 1   EVOLUTIONARY OPERATIONS—GRAPHICAL PRESENTATION OF TREAT-
MENTS

ent time and temperature as the central point and variations in all direction.

This naturally leads to a design where the number of time-temperature
variations tried are five: (1) current procedure; (2) higher temperature and
shorter time; (3) lower temperature and shorter time; (4) lower temperature
and longer time; and (5) higher temperature and longer time. The deviations from the central point should, at least in the beginning, be small
enough that they do not affect the end product to the extent that it might
become unsalable. Since variations among these treatments are small, it is
not likely that a significant difference can be obtained by running through
these 5 treatments only once. If, after these 5 treatments (referred to as one
cycle) are replicated, say for 5 or 6 cycles, there is still no significant
difference in response in terms of yield or quality, it would then be appropriate to increase the size of the increments in one or both directions. The
process would then be repeated until a significant difference emerged. Each
time enough cycles are completed to justify changing the treatments, another *phase* of the EVOP procedure begins.

If it should be found that treatment X_1 (the current process) is significantly better than any variation in any direction, then this EVOP study
could still be continued, but less often. If, however, one of the peripheral
treatments ($X_{2,3,4,\text{or }5}$) should be found to be significantly superior to the

central treatment (X_1), then the adjustments would be made in the equipment to proceed with the new "best" time and temperature as the central treatment (Y_1), the previous central treatment serving as one of four new peripheral treatments ($Y_{2,3,4,or\ 5}$). In this way, therefore, there would be a continuous evolutionary probing in all directions for a best treatment under possibly changing conditions.

The stream of information concerning the products from the various manufacturing conditions is summarized on an *information board* prominently displayed in the plant so that interested personnel can follow the progress of the operation. The entries shown on the Information Board (Fig. 16.2) are (1) the number of the present phase, (2) the number of the last cycle completed, (3) the average response observed at each set of conditions up to the conclusion of the last cycle, (4) the "error limits" for these averages (based on a prior estimate of standard deviation), (5) the effects with their associate error limits, (6) the standard deviation calculated from the data of this particular phase, and (7) the prior estimate of the standard deviation available from previous work. The "effects" are the usual main effects and 2-factor interactions appropriate for a 2-level factorial experiment, together with the "change in mean" effect. This latter quantity measures the difference between the average yield at the center conditions and the average yield at the conditions being run in this phase.

The entries on the Information Board are changed at the end of each cycle. The calculations required to derive these data are based on simple statistical procedures and are usually performed on a standard worksheet.

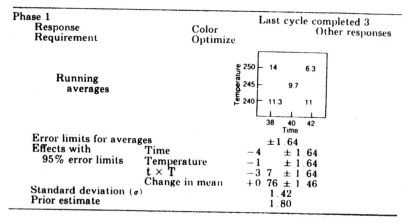

FIG. 16.2.   INFORMATION BOARD FOR CORN COLOR DATA

An EVOP Example Using Standard Procedures

In order to demonstrate the use of EVOP, we will use a corn color example. Within an EVOP framework, the goal is to see if color can be significantly improved by manipulating, very slightly, the process variables of time and temperature.

In this hypothetical case, the EVOP committee, starting with a current process time and temperature of 40 min at 245°F (118.3°C), (treatment 1), selects the other treatment values, (2) 38 min at 240°F (115.6°C), (3) 42 min at 250°F (121.1°C), (4) 42 min at 240°F (115.6°C), and (5) 38 min at 250°F (121.1°C) (Fig. 16.3). Obviously, a prior decision must have been reached to assure the adequacy of all 5 processing treatments from the spoilage hazard standpoint.

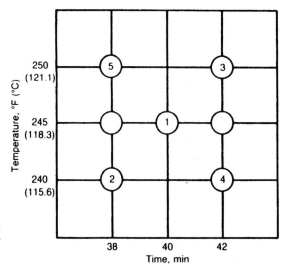

FIG. 16.3: A GRAPHIC PRESENTATION OF A CYCLE OF TREATMENTS

Cycle One.—The calculations required to determine if we obtain significant results are normally performed on a *calculation worksheet*. After completing one cycle (going through all 5 treatments one time), the worksheet might look like Table 16.1.

For the values of color the EVOP committee has chosen the minimum color for the USDA grade of fancy as having a value of 10. This may be an arbitrary value or in the case of some responses such as yield, the values may be expressed as percentage yield. Since no previous observations are available, the previous cycle sum (line I), the previous cycle average (line II), and the difference (line IV) cannot be filled in. The values given to the color of the corn from each of the 5 sets of operating conditions are recorded in line III. These evaluations are usually made by quality control personnel. In the line labeled "new sums" and "new averages" are recorded the sum and average of all the observations recorded at each of the operating conditions up to and including the cycle being considered. In the present case where we are reviewing the situation at the end of the first cycle, these sums and averages are simply the observations themselves.

TABLE 16.1. WORKSHEET FOR DATA FROM CYCLE ONE: 2-VARIABLE EVOP PROGRAM
Cycle n = 1
Response—Color

Calculation of Averages						Calculations of "S"
Operating conditions	(1)	(2)	(3)	(4)	(5)	
(I) Previous cycle sum						Previous sum S =
(II) Previous cycle average						Previous average S =
(III) New observations	10	11	8	12	14	New S = R × $f_{k,n}$ =
(IV) Differences (II)−(III)						Range =
(V) New sums	10	11	8	12	14	New sum S =
(VI) New averages	10	11	8	12	14	New average S =

Calculation of Effects	Calculation of Error Limits
Time effect = $\frac{1}{2}(\bar{y}_3 + \bar{y}_4 - \bar{y}_2 - \bar{y}_5) = -2.5$	New average = $2\sigma/\sqrt{n} = \pm 3.6$
Temperature effect =	
$\frac{1}{2}(\bar{y}_3 + \bar{y}_5 - \bar{y}_2 - \bar{y}_4) = -0.5$	New effects = $2\sigma/\sqrt{n} = \pm 3.6$
$t \times T$ effect = $\frac{1}{2}(\bar{y}_2 + \bar{y}_3 - \bar{y}_4 - \bar{y}_5) = -3.5$	Change in mean =
Change in mean =	$1.78\sigma/\sqrt{n} = \pm 3.2$
$\frac{1}{5}(\bar{y}_2 + \bar{y}_3 + \bar{y}_4 + \bar{y}_5 - 4\bar{y}_1) = 1$	Prior estimate $\sigma = 1.8$

The standard deviation cannot be estimated from the data of this phase since only one observation is available at each of the operating conditions. When a prior estimate of the standard deviation is available, this can be used in calculating provisional error limits. If one is not available, an intelligent estimate must be made based on the nature of the product.

In the calculation of effects, the time effect is calculated by taking one-half of the value of the responses at the 2 longer times (treatments 3 and 4 at 42 min) minus the response at the 2 shorter times (treatments 2 and 5 at 38 min). The temperature effect is calculated by taking one-half of the value of the response at the 2 higher temperatures [treatments 3 and 5 at 250°F (121.1°C)] minus the response at the 2 lower temperatures [treatments 2 and 4 at 240°F (115.6°C)]. The combined effect of time and temperature is calculated by taking one-half of the value at treatments 2 and 3 minus one-half of the value at treatments 4 and 5. The change in mean effects measures the difference between the average yield at the center condition and the average yield at all the conditions being run in this phase. In this example, it is calculated by taking $\frac{1}{5}$ of the value of the responses at treatments 2, 3, 4, and 5 minus 4 times the value of 1.

The calculation of the error limits is based upon the standard deviation of the phase. Since this is the first cycle of the first phase we have no figures upon which to determine a standard deviation so we use a prior estimate of standard deviation based on previous work. With this value we calculate the error limits for the new averages, for the new effects, and for the new change in means.

Cycle Two.—At cycle 2 (Table 16.2), we begin to obtain information from the data themselves concerning the value of the standard deviation relevant to this phase. The data from lines V and VI on the calculation worksheet of the previous cycle are copied in lines I and II of the worksheet for the present cycle. The observations recorded during the present cycle are listed in line III. The difference between the quantities listed on lines II and III is written in line IV. These are the differences with appropriate signs attached between the new observations and the averages over previous cycles taken at each set of operating conditions. The new sum and new average for each set of conditions are recorded on lines V and VI. The effects are calculated as given on the worksheet using the new averages computed at each of the sets of operating conditions.

TABLE 16.2.   WORKSHEET FOR DATA FROM CYCLE TWO: 2-VARIABLE EVOP PROGRAM CALCULATION WORKSHEET
Cycle n = 2
Response—Color

Calculation of Averages						Calculations of "S"
Operating conditions	(1)	(2)	(3)	(4)	(5)	
(I) Previous cycle sum	10	11	8	12	14	Previous sum S =
(II) Previous cycle average	10	11	8	12	14	Previous average S =
(III) New observations	8	12	7	10	13	New S = R $\times$ $f_{k,n}$ = 0.9
(IV) Differences (II)−(III)	2	−1	1	2	1	Range = 3
(V) New sums	18	23	15	22	27	New sum S = 0.9
(VI) New averages	9	11.5	7.5	11	13.5	New average S = 0.9

Calculation of Effects	Calculation of Error Limits
Time effect = $\tfrac{1}{2}(\bar{y}_3 + \bar{y}_4 - \bar{y}_2 - \bar{y}_5) = -3.25$	New average = $2\sigma/\sqrt{n}$ = ±2.5
Temperature effect =	
$\tfrac{1}{2}(\bar{y}_3 + \bar{y}_5 - \bar{y}_2 - \bar{y}_4) = -0.75$	New effects = $2\sigma/\sqrt{n}$ = ±2.5
t × T effect = $\tfrac{1}{2}(\bar{y}_2 + \bar{y}_3 - \bar{y}_4 - \bar{y}_5) = -2.75$	Change in mean =
	$1.78\sigma/\sqrt{n}$ = ±2.2
Change in mean =	
$\tfrac{1}{5}(\bar{y}_2 + \bar{y}_3 + \bar{y}_4 + \bar{y}_5 - 4\bar{y}_1) = 1.50$	

After the second cycle it is possible to obtain an estimate of the standard deviation of the individual observations (and therefore of the error limits for the averages and the effects) from the data themselves. The largest and smallest differences recorded in line IV are underlined. The range of the differences is the difference between these underlined values. This is entered at the right hand end of line IV. The range in line IV is multiplied by the factor $f_{k,n}$, obtained from a table of constants found in Table 16.3. Since at cycle 2 no previous estimate of the standard deviation is available from the data, the items "previous sum S" and "previous average S" are blank and the entries for "new sum S" and "new average S" are identical with that in line III.

TABLE 16.3.   VALUES OF CONSTANT $f_{k,n}$

Number of Cycles = n	k = Number of Sets of Conditions in Block								
	2	3	4	5	6	7	8	9	10
2	0.63	0.42	0.34	0.30	0.28	0.26	0.25	0.24	0.23
3	0.72	0.48	0.40	0.35	0.32	0.30	0.29	0.27	0.26
4	0.77	0.51	0.42	0.37	0.34	0.32	0.30	0.29	0.28
5	0.79	0.53	0.43	0.38	0.35	0.33	0.31	0.30	0.29
6	0.81	0.54	0.44	0.39	0.36	0.34	0.32	0.31	0.30
7	0.82	0.55	0.45	0.40	0.37	0.34	0.33	0.31	0.30
8	0.83	0.55	0.45	0.40	0.37	0.35	0.33	0.31	0.30
9	0.84	0.56	0.46	0.40	0.37	0.35	0.33	0.32	0.31
10	0.84	0.56	0.46	0.41	0.37	0.35	0.33	0.32	0.31
11	0.84	0.56	0.46	0.41	0.38	0.35	0.33	0.32	0.31
12	0.85	0.57	0.47	0.41	0.38	0.35	0.34	0.32	0.31
13	0.85	0.57	0.47	0.41	0.38	0.36	0.34	0.32	0.31
14	0.85	0.57	0.47	0.41	0.38	0.36	0.34	0.32	0.31
15	0.86	0.57	0.47	0.42	0.38	0.36	0.34	0.33	0.31
16	0.86	0.57	0.47	0.42	0.38	0.36	0.34	0.33	0.32
17	0.86	0.57	0.47	0.42	0.38	0.36	0.34	0.33	0.32
18	0.86	0.57	0.47	0.42	0.38	0.36	0.34	0.33	0.32
19	0.86	0.58	0.47	0.42	0.38	0.36	0.34	0.33	0.32
20	0.86	0.58	0.47	0.42	0.38	0.36	0.34	0.33	0.32

The estimate of the standard deviation after the second cycle is not very reliable and the prior estimate would normally be used in calculating the error limits. With k = 5 (5 sets of conditions) we would usually wait until 3 cycles have been completed before the value for standard deviation obtained from the data would be employed.

Cycle Three.—All the entries in the calculation worksheet can be filled at the third cycle (Table 16.4). The value of "new S" obtained from the current cycle is added to the "previous sum S" and the result is recorded as "new sum S." This quantity is then divided by (n − 1) to give a "new average S." The error limits for the averages and the effects are obtained by direct substitution of "new averages" for standard deviation in the equations in the lower right portion of the calculation worksheet. The new averages, estimated effects, and their error limits are recorded on the Information Board (Fig. 16.2).

Conclusions.—At this point in our example the plant manager can get a good idea of just how his process is progressing and in which direction he should move his variables. From the Information Board it seems that we are getting a decidedly better color response at a treatment of 250°F (121.1°C) for 38 min than at our current treatment of 245°F (118.3°C) for 40 min. Based on this conclusion the plant manager may be justified in making treatment 5 [250°F (121.1°C) for 38 min] the center treatment and deriving a new set of values for the other 4 sets of conditions. A new set for a second phase may be (1) 250°F (121.1°C) for 38 min, (2) 245°F (118.3°C) for 36 min, (3) 255°F (123.9°C) for 40 min, (4) 245°F (118.3°C) for 40 min, and (5) 255°F (123.9°C) for 36 min. Or the plant manager may decide that not enough cycles have been run to justify a change, or he might ask for suggestions from the EVOP committee.

TABLE 16.4.   WORKSHEET FOR DATA FROM CYCLE THREE:
2-VARIABLE EVOP PROGRAM CALCULATION WORKSHEET
Cycle n = 3
Response—Color

Calculation of Averages						Calculations of "S"
Operating conditions	(1)	(2)	(3)	(4)	(5)	
(I) Previous cycle sum	18	23	15	22	27	Previous sum S = 0.90
(II) Previous cycle average	9	11.5	7.5	11	13.5	Previous average S = 0.90
(III) New observations	11	11	4	11	15	New S = R × $f_{k,n}$ = 1.93
(IV) Differences (II)−(III)	−2	0.5	3.5	0.0	−1.5	Range = 5.50
(V) New sums	29	34	19	33	42	New sum S = 2.83
(VI) New averages	9.7	11.3	6.3	11	14	New average S = 1.42

Calculation of Effects	Calculation of Error Limits

Time effect = $\frac{1}{2}(\bar{y}_3 + \bar{y}_4 - \bar{y}_2 - \bar{y}_5)$ = −4   New average = $2\sigma/\sqrt{n}$ = ±1.64

Temperature effect =

$\frac{1}{2}(\bar{y}_3 + \bar{y}_5 - \bar{y}_2 - \bar{y}_4)$ = −1   New effects = $2\sigma/\sqrt{n}$ = ±1.64

t × Teffect = $\frac{1}{2}(\bar{y}_2 + \bar{y}_3 - \bar{y}_4 - \bar{y}_5)$ = −3.7   Change in mean =

$1.78\sigma/\sqrt{n}$ = ±1.46

Change in mean =

$\frac{1}{5}(\bar{y}_2 + \bar{y}_3 + \bar{y}_4 + \bar{y}_5 - 4\bar{y}_1)$ = 0.76

An EVOP Example Using Modified Procedures

Kramer (1964) has modified Box's EVOP procedure by using ranks instead of actual values for the averages. This has further simplified the EVOP procedure by eliminating the calculations necessary to determine the standard deviation and increasing the ease by which to determine significance of results for the different phases. This method gives the plant manager the relative values for each of the treatments under observation and allows him to make his decision as to which direction to carry on his treatments.

To illustrate this method, an example will be used of the blanching operation on green beans for canning. The variables we are considering are time and temperature of a water blanch. Our purpose is to optimize quality by finding the particular time and temperature of blanch which would minimize the degree of sloughing of the finished product. We select as our central treatment the current process, which happens to be a 3 min blanch at 190°F (87.8°C) (X_1). Our 4 peripheral treatments are: 2 min at 200°F (93.3°C) (X_2), 2 min at 180°F (82.2°C) (X_3), 4 min at 180°F (82.2°C) (X_4), and 4 min at 200°F (93.3°C) (X_5). We have selected these treatments because we assume that such variations, although they may have some effect on the degree of sloughing, will still not cause an extreme effect that will make the product unusable.

We now begin collecting our data by picking up a few samples of the finished product blanched 3 min at 190°F (87.8°C), examining the product for degree of sloughing, and recording that value. We then adjust the blancher so that the beans are blanched at a higher temperature but for a shorter time (treatment X_2). When we are satisfied that the beans in the finished product have undergone the new treatment, we collect a few more samples of the finished product and examine these for degree of sloughing, and record the value again. We continue with this same procedure, changing the blanching conditions to cover all 5 treatments, and record them as shown in Table 16.5. We now convert the 5 values in terms of number of sloughed units to ranks, in this case ranking the treatment with the smallest number as 1 and the treatment with the largest number of sloughed units as 5.

TABLE 16.5.   EFFECT OF TIME AND TEMPERATURE OF BLANCH ON THE NUMBER OF SLOUGHED GREEN BEANS

	Blanching Treatments				
	X_1 3 min 190°F (87.8°C)	X_2 2 min 200°F (93.3°C)	X_3 2 min 180°F (82.2°C)	X_4 4 min 180°F (82.2°C)	X_5 4 min 200°F (93.3°C)
Replication	Count Rank	Count Rank	Count Rank	Count Rank	Count Rank
1	7 – 1	8 – 2	10 – 3	14 – 5	13 – 4
2	9 – 3	6 – 1	7 – 2	12 – 4	13 – 5
Σ, replications 1 + 2	4	3	5	9	9
3	9 – 3.5	7 – 1	8 – 2	9 – 3.5	11 – 5
Σ, replications 1 + 2 + 3	7.5	4	7	12.5	14

At this point it appears that the central treatment (X_1), that is, the current practice, is the best treatment. However, it is impossible for us to prove statistical significance among 5 treatments with only this 1 replication, so that we must repeat the entire series at least once more. In the second replication we find that treatment X_2, the 2-min blanch at 200°F (93.3°C), appears to be best so it is ranked first (Table 16.5). Having 2 replications, we may now sum the ranks for each treatment to see if we have a significantly best treatment at this point. With 5 treatments and 2 replications, any rank sum that is less than 3 may be considered significantly different (Appendix Table I.1). We seem to be approaching this with treatment X_2 but we have not as yet adequate proof that treatment X_2 is significantly better than the others. We therefore erase the rank sums at this point and continue with another replication of our 5 treatments. Actually, we could erase all the data except the rank sums and continue with replication 3. After obtaining sloughing counts for all 5 treatments for the third cycle, we add these rank values to obtain new rank sums of all 3 replications (Table 16.5). For 5 treatments and 3 replications, we may consider any rank sum less than 5 or greater than 13 to be significant (Appendix Table I.1).

We may therefore conclude that treatment X_2, that is, 2 min at 200°F (93.3°C) is the best treatment, and so we are justified in changing our procedure in the direction of a shorter but hotter blanch.

Thus treatment number 2 becomes the new central treatment (Y_1) and, for the time being, the recommended process. The former current process (X_1) now becomes one of the peripheral treatments (Y_4), and the 3 other peripheral treatments can then be set at 1 min at 210°F (98.9°C) (Y_3), 1 min at 190°F (87.8°C) (Y_3), and 3 min at 210°F (98.9°C) (Y_3). The EVOP procedure may then continue with these 5 new treatments until one of them shows a significant improvement over the central treatment.

Having moved our central treatment from X_1 to Y_1, we may have uncovered the first step in a "slope of steepest ascent" toward a shorter and hotter process. If, after further observations of the Y treatments, we should find treatment Y_2 superior, then the direction of the steepest ascent would be unmistakable (Fig. 16.4).

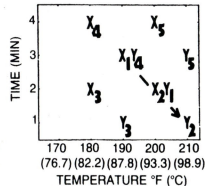

FIG. 16.4. GRAPHIC PRESENTATION OF 5 EVOP TREATMENTS ILLUSTRATING THE EMERGENCE OF A SLOPE OF STEEPEST ASCENT

SUMMARY

EVOP has been demonstrated to be a valuable decision aid in day to day plant operations. It is based on a simple 2-level factorial arrangement. It uses range information to estimate the variance. As with the regular ANOV model, the power of the test increases as the number of cycles increases (i.e., as the number of replicates increases).

Acceptance of EVOP by production personnel, in general, is quite good. The procedure requires only minimal departures from normal operating conditions. In addition, there is immediate feedback indicating whether or not improvements are being achieved. The use of a worksheet, similar to that shown in Table 16.6, does not impose an inordinate computational burden on production personnel.

TABLE 16.6. 2-VARIABLE EVOLUTIONARY OPERATION PROGRAM CALCULATION WORKSHEET

Project _____
Phase _____
Date _____

$$\text{CYCLE } n =$$

Response _____

Calculation of Averages

Operating Conditions	(1)	(2)	(3)	(4)	(5)
(a) Previous cycle sum					
(b) Previous cycle average					
(c) New observations					
(d) Differences (b) minus (c)					
(e) New sums					
(f) New average $\bar{y} = \dfrac{(e)}{n}$					

Calculation of Standard Deviation

(g) Previous sum s =

(h) Previous average s =

(i) New s = Range $\times f_{k,n}$ =

(j) Range (= d) =

(k) New sum s =

(l) New average s = $\dfrac{(k)}{(n-1)}$

Calculation of Effects

_____ effect $= \frac{1}{2}(\bar{y}_3 + \bar{y}_4 - \bar{y}_2 - \bar{y}_5) =$

_____ effect $= \frac{1}{2}(\bar{y}_3 + \bar{y}_5 - \bar{y}_2 - \bar{y}_4) =$

_____ effect $= \frac{1}{2}(\bar{y}_2 + \bar{y}_3 - \bar{y}_4 - \bar{y}_5) =$

Change in mean effect $= \frac{1}{5}(\bar{y}_2 + \bar{y}_3 + \bar{y}_4 + \bar{y}_5 - 4\bar{y}_1) =$

Calculation of Error Limits

For new average $= \dfrac{2}{\sqrt{n}} s = \pm$

For new effects $= \dfrac{2}{\sqrt{n}} s = \pm$

For change in mean $= \dfrac{1.78}{\sqrt{n}} s = \pm$

Two words of caution should be borne in mind. First, EVOP works best in continuous production processes. It has been applied with only limited success in "job shop situations" because of the difficulty of achieving a true replicate. Second, optimizing some characteristic (color, volume, etc.) may be achieved at a cost which can not be justified. Consequently, it is necessary for management to monitor results to ensure that the objectives achieved through EVOP contribute to the overall goals of management.

PROBLEMS

16.1. Cake Volumes I
Given the following production runs in a bakery where EVOP is being used to maximize cake volume:

1	2	3	4	5
325°F	320°F	330°F	320°F	330°F
162.8°C	160°C	165.6°C	160°C	165.6°C
25 min	23 min	27 min	27 min	23 min
50	47	50	48	50
48	48	52	48	51
50	47	52	47	49
52	49	49	49	50
49	46	53	50	49
50	49	52	49	50
51	50	51	50	51

(a) Determine at which cycle a change in the production process would be made.
(b) What would be the design of the new EVOP pattern as a result of your decision in step (a)?

16.2. Cake Volumes II
Using the data for problem 16.1, convert these observations to ranks. Use Kramer's modified EVOP technique to evaluate this production process.

16.3. Cake Volumes III
Ignoring the center point observation (i.e., point 1—325°F (162.8° C) and 25 min), evaluate the data as the results of an experiment utilizing a 2 × 2 factorial design. How do these results compare with the conclusions that you reached when you analyzed the data for 16.1 within an EVOP framework?

REFERENCES

Suggested Further Reading

For a detailed exposition of the use of EVOP see:
BOX, G.E.P. and DRAPER, N.R. 1969. Evolutionary Operation. John Wiley & Sons, New York.

Selected Other Readings

BINGHAM, R.S., JR.   1963.   Try EVOP for systematic process improvement. Ind. Qual. Control 20 (3) 17–23.

BOX, G.E.P.   1954.   The exploration and exploitation of response surfaces: some general considerations and examples. Biometrics 10, 16–60.

BOX, G.E.P.   1957.   Evolutionary operation: A method for increasing industrial productivity. Appl. Stat. 6 (2) 3–23.

BOX, G.E.P. and HUNTER, J.S.   1959.   Condensed calculations for evolutionary operation programs. Technometrics 1 (1) 77–95.

BOX, G.E.P. and WILSON, K.B.   1951.   On the experimental attainment of optimal conditions. J. R. Stat. Soc. B13, 1–38.

DAVIES, O.L.   1967.   The Design and Analysis of Industrial Experiments. Hafner Publishing Co., New York.

DUNCAN, A.J.   1959.   Mapping response surfaces and determination of optimum conditions. In Quality Control and Industrial Statistics.   Richard D. Irwin, Homewood, Ill.

FERRELL, E.B.   1956.   The Control Chart—Modifications and Extension. Trans. Mid-Atlantic Conf., New York, 1956. Am. Soc. Qual. Control, New York.

FOX, M.   1956.   EVOP (Evolutionary Operation)—A practical application. Md. Processors Rep. 11 (1) 1–8.

GRANT, E.L.   1946.   Statistical Quality Control: Some fundamental Statistical Concepts, 1st Edition.   McGraw-Hill Book Co., New York.

HERMANSON, H.P.   1965.   Maximization of potato yield under constraint. Agron. J. 57, 210.

KISSELL, L.T.   1967.   Optimization of white layer cake formulations by a multiple-factor experimental design. Cereal Chem. 44, 253–268.

KRAMER, A.   1963.   Revised tables for determining significance of differences. Food Technol. 17 (12) 124–125.

KRAMER, A.   1964.   The effective use of operations research and EVOP in quality control. Food Technol. 19 (1) 37–39.

KRAMER, A. and TWIGG, B.A.   1970.   Quality Control for the Food Industry, 3rd Edition, Vol. 1.   AVI Publishing Co., Westport, Conn.

KRAMER, A. and TWIGG, B.A.   1973.   Quality Control for the Food Industry, 3rd Edition, Vol. 2.   AVI Publishing Co., Westport, Conn.

MOORE, D.D. et al.   1957.   An investigation of some of the relationships between copper, iron, molybdenum in the growth and nutrition of lettuce. II. Response surfaces of growth and accumulation of Cu and Fe. Proc. Soil Sci. Soc. Am. 21, 65.

PEARSON, A.M., BATEN, W.D., GOEMBEL, A.J., and SPOONER, M.E. 1962.   Application of surface-response methodology to predicting optimum levels of salt and sugar in cured ham. Food Technol. *16*, 137–138.

SMITH, H. and ROSE, A.   1963.   Subjective responses in process investigation. Ind. Eng. Chem. *55*, 25.

WILDE, D.J.   1964.   Optimum Seeking Methods.   Prentice-Hall, Englewood Cliffs, N.J.

17

Index Numbers

Index numbers are a common factor of everyday life. The Consumer Price Index (CPI), the Producer Price Index (PPI), which was formerly known as the Wholesale Price Index (WPI), Agricultural Parity Prices, and the Dow Jones Industrial Average are all well-known terms. Each is an index in the sense that each is a statistic which summarizes a host of economic activities. For example, although there are thousands of consumer items which could be purchased in the United States, the Consumer Price Index is constructed using a "market basket" of approximately 250 categories. In spite of the fact that an individual might feel intuitively (or in his pocketbook) what is happening to the general price level, he could not realistically monitor the movement of all prices simultaneously. The Consumer Price Index is an attempt to distill all consumer price movements into a single value.

The strength of an index is also its weakness. Many critics of index numbers argue that a single value cannot be representative of several hundred statistical series. In other words, the CPI does not really reflect the changing prices for 250 goods and services since some prices are rising, others are falling, and still others are constant. There is considerable merit to this argument. However, the use of index numbers is so widespread that no amount of criticism will eliminate their use. Furthermore, there is no need to eliminate the use of index numbers, but it is important to understand how index numbers are constructed in order to use and interpret them correctly.

INDEX FORMULAS

The two most common formulas for constructing indices are:

Laspeyres Formula

$$L = \frac{\Sigma p_c q_b}{\Sigma p_b q_b} \cdot 100$$

and

264

Paasche Formula

$$P = \frac{\Sigma p_c \, q_c}{\Sigma p_b \, q_c} \cdot 100$$

where L = Laspeyres Index
    P = Paasche Index
    p_c = Price in the current period
    p_b = Price in the base period
    q_c = Quantity in the current period
    q_b = Quantity in the base period

Both of these formulas relate current prices to prices in some base period. The Laspeyres Index uses quantities defined for the base period as weights. The Paasche Index uses quantities in the current period as weights. Since the determination of quantities is often a time consuming and expensive process, most indices use the Laspeyres formula.

Because the Laspeyres Index uses quantities from a base period, it overstates price increases and understates price decreases. For example, if the price of a single item increases dramatically, we would expect consumers to purchase less of this good. However, since the Laspeyres Index holds the base quantity constant, no substitution of cheaper goods is accounted for and the Laspeyres Index overstates the impact of the price rise. In a similar manner, when the price of a good falls, we expect consumers to purchase more of it. Since the Laspeyres Index holds base quantities constant, it understates the impact of a price decline.

Because it uses current quantities for weights, the Paasche Index understates the impact of price increases and overstates the effect of falling prices.

In order to overcome these difficulties, Drobisch suggested in 1871 that a simple average of the two indices be used:

$$D = \frac{L + P}{2}$$

In 1920, Irving Fisher defined the Ideal Index as the geometric mean of the two indices:

Fisher's Ideal Index $= (L \times P)^{1/2}$

Since few agencies or individuals are willing to incur the expense of collecting current quantity data in order to calculate a Paasche Index, Drobisch's Index and Fisher's Ideal Index are almost never encountered in practice. In addition, it should be noted that the values obtained are usually very close, whether the Laspeyres Index or Paasche Index is used, if the weights are composed of several hundred items, as is commonly the situation.

LOGICAL CRITERIA OF INDEX NUMBERS

Statisticians have from time to time addressed the question concerning the mathematical properties which an index number should possess. In order to discuss these properties, we need to introduce three indices that are relatively uncommon.

Instead of using quantities as weights in order to examine changing prices, we could use prices as weights in order to analyze changing quantities. Such a *quantity index* using Laspeyres formula would be:

$$Q_L = \frac{\Sigma p_b \, q_c}{\Sigma p_b \, q_b} \times 100$$

For the Paasche formula, the index would be:

$$Q_P = \frac{\Sigma p_c \, q_c}{\Sigma p_c \, q_b} \times 100$$

The third index that we need is the value index.

$$V = \frac{\Sigma p_c \, q_c}{\Sigma p_b \, q_b} \times 100$$

That is, the value index (V) is simply the ratio of the total value of goods in the current period to the total value in the base period (times 100).

The *factor reversal test* requires that the product of a price index and its corresponding quantity index equal its value index. Neither the Laspeyres Index nor the Paasche Index will meet this test. However, Fisher's Ideal Index will.

The *time reversal test* requires that the interchange of the current and base periods yield the reciprocal of the index. In other words, this test requires that for the Laspeyres Index:

$$\frac{\Sigma p_c \, q_b}{\Sigma p_b \, q_b} \times 100 = \frac{1}{\frac{\Sigma p_b \, q_c}{\Sigma p_c \, q_c} \times 100}$$

which is clearly untrue.

Again, neither the Laspeyres Index nor the Paasche Index will meet this test. Fisher's Ideal Index does meet the *time reversal test.*

The primary reason for showing that the common indices do not possess the mathematical properties that one might expect is to underscore the need to understand how index numbers are constructed in practice in order to minimize the misuse of index numbers.

METHODS OF CONSTRUCTION

The appropriate construction of an index depends primarily on its purpose. We may be interested in an index of productivity of workers in various plants. Alternatively, we may be interested in an index of marketing activity. Most commonly, we are interested in interpreting and using an index calculated by someone else.

Choice of Base Period

Ordinarily, an index is used to compare movement in a statistical series or a group of series over time. Consequently, the base period is chosen to represent a period of normalcy. As conditions change over time, the base period may be shifted to a more recent date. For example, most statistical indices in the United States use 1967 as the base year (i.e., 1967 = 100). Previously, many of these indices used the 3 year average 1957–1959 as the base period (i.e., 1957–1959 = 100). Some indices are tied by law to a specific base period (e.g., agricultural parity prices must use the period 1910–1914 as their base).

In general, a base period should be chosen to reflect average or normal conditions. In addition, the base period should be recent enough that the weights used in the calculation of the index and the interpretations derived from the index are meaningful.

Selection of Weights

Occasionally, the selection of weights is a simple and straightforward process. For example, when developing productivity indices for a single firm (even when the firm is a multiplant, multiproduct organization), it may be simplest to include the output of all products produced. However, in general, it is not possible to include all items and a selection must be made. The question becomes one of selecting weights which represent 70, 80, 90%, or some other fraction of the total for which the index is being developed. The Consumer Price Index uses weights which represent more than 90% of the expenditures of an urban family. In contrast, the Dow Jones Industrial Average uses weights which represent less than 10% of the stocks traded on the New York Stock Exchange.

The selection of weights requires considerable judgment and experience. When constructing an index for internal use, it may be desirable to consider different sets of weights evaluating the gain in information against the increased cost of calculating the index as more items are used for weights. When using a published index, it is usually desirable to determine what weights are used since this may influence the interpretation of the index.

Calculation and Use of an Index

The calculation of an index requires the definition of a base period, the determination of the appropriate weights, and the recording of prices. Tables 17.1, 17.2, and 17.3 indicate the nature of these steps.

The base period has been chosen as 1967. That is, 1967 = 100.

Table 17.1 shows the simplified (and somewhat arbitrary) weights chosen for this example. In practice, the weights are determined by a careful survey

TABLE 17.1.   WEIGHTS FOR A SIMPLIFIED CONSUMER PRICE INDEX BASED ON AN URBAN FAMILY OF 4

Food	
meat	400 lb (181.4 kg)
dairy (milk)	100 gal. (375.5 liters)
vegetables	800 lb (362.9 kg)
Shelter (5 room apartment)	1 unit
Transportation	
taxi rides	6 rides
gasoline for family car	700 gal. (2.65 kl)
Medical	
visits to the family doctor	10 visits
Clothing	
dresses	5 dresses
men's suits	1 suit

TABLE 17.2.   PRICES (DOLLARS) FOR A SIMPLIFIED CONSUMER PRICE INDEX

Item	Year			
	1967	1978	1979	1980
Meat ($/lb)	1.00	1.80	2.00	2.20
Milk ($/gal.)	1.00	1.80	2.00	2.00
Vegetables ($/lb)	0.15	0.30	0.30	0.30
Rent ($/month)	200.00	390.00	400.00	410.00
Taxi ($/ride)	2.00	4.00	4.00	5.00
Gasoline ($/gal.)	0.40	0.70	0.90	1.00
Doctors ($/visit)	8.00	15.00	15.00	16.00
Dresses ($/dress)	30.00	40.00	45.00	50.00
Suits ($/suit)	100.00	170.00	175.00	180.00

TABLE 17.3.   CALCULATION OF AN INDEX FOR 1967 AND 1979

Item	Weight	1967 Price $	Value	Weight	1979 Price $	Value
Meat	400	1.00	400.00	400	2.00	800.00
Milk	100	1.00	100.00	100	2.00	200.00
Vegetables	800	0.15	120.00	800	0.30	240.00
Rent	1	200.00	2400.00	1	400.00	4800.00
Taxi	6	2.00	12.00	6	4.00	24.00
Gasoline	700	0.40	280.00	700	0.90	630.00
Doctor	10	8.00	80.00	10	15.00	150.00
Dresses	5	30.00	150.00	5	45.00	225.00
Suits	1	100.00	100.00	1	175.00	175.00
Total			3642.00			7244.00

of consumer behavior. Furthermore, the weights usually represent a survey that is more recent than the base period. The market basket may be changed from time to time even though the base period is left unchanged. In this case, the weights represent the consumer market basket for 1975. A common price index normally uses annual consumption data for its weights. Table 17.1 uses weights which might represent reasonable annual quantities for an urban family of 4.

Table 17.2 shows the market prices collected on these items for the years 1967, 1978, 1979, and 1980. As can be seen in Table 17.2, prices are changing at different rates. The purpose of an index is to summarize these varying movements into a single value that can be interpreted.

Table 17.3 shows the calculations necessary to calculate the price indices for 1967 and 1979.

Using Laspeyres formula, the index for 1967 is:

$$L_{67,67} = \frac{\Sigma p_c q_b}{\Sigma p_b q_b} \cdot 100$$

$$= \frac{3642.00}{3642.00} \cdot 100$$

$$= 100.0$$

For 1979 the index is:

$$L_{67,79} = \frac{\Sigma p_c q_b}{\Sigma p_b q_b} \cdot 100$$

$$= \frac{7244.00}{3642.00} \cdot 100$$

$$= 198.9$$

In a similar manner the indices for 1978 and 1980 can be calculated:

$$L_{67,78} = \frac{6854.00}{3642.00} \cdot 100$$

$$= 188.2$$

$$L_{67,80} = \frac{7560.00}{3642.00} \cdot 100$$

$$= 207.6$$

Once such an index is constructed, it can be used to compare items over time. For example, a family whose income is $10,000 in 1967 would need $19,890 (10,000 $\times$ 1.989) in 1979 just to purchase the same bundle of goods and services. That is, it would require a near doubling of money income in order to hold real income constant.

In this way, index numbers help us cut through a maze of data in order to grasp the nuggets of information that lie within.

COMMON INDICES

The most common index is the Consumer Price Index (CPI) which is calculated by the Bureau of Labor Statistics, U.S. Department of Labor. First established in 1919, it was called the Cost of Living Index until 1945. However, the CPI is not a cost of living index. For example, it excludes income taxes and social security taxes. Although research is being conducted on cost of living indices (Bannerjee 1975), it is unlikely that a true cost of living index will be constructed due to the cost of data collection and definitional problems.

The CPI is revised periodically. Table 17.4 lists the reference periods that have been used since its inception. Currently, the CPI has two primary series: the traditional index for wage earners and clerical workers, and a new series that covers all urban consumers. In addition, the CPI is partitioned into subcomponents (all items, food, housing, apparel, and upkeep, etc.).

TABLE 17.4.   CONSUMER PRICE INDEX REFERENCE BASE AND PERIODS OF USE

Reference Bases (= 100)	Periods in Use	
	From	Through
1913	1913	September 1935
1923 – 1925	October 1935	June 1940
1935 – 1939	July 1940	1952
1947 – 1949	1953	1961
1957 – 1959	1962	1970
1967	1971	To date

Prior to 1978, the CPI was constructed using a market basket of approximately 400 items with very specific definitions. Currently, approximately 250 categories are used to achieve a slightly broader coverage of goods and services. For example, the old market basket item "Vitamin D, Grade A Homogenized milk in half-gallon containers" has been replaced with the category "whole fresh milk" which enables the predominant form of milk to be priced rather than a specific item which may be sold on a sporadic basis.

The current market basket of goods and services is based on a survey of consumer expenditures during 1972–1973, while the base of the index is still 1967 = 100 for prices.

One of the most important uses of a price index is to convert observed (or nominal) prices into deflated (or real) prices in order to eliminate the distortion of inflation. Table 17.5 reports the observed price and per capita consumption of broilers in the United States for the 10 year period 1966–1975. During that period, both per capita consumption and the observed price increased. This information is plotted in Fig. 17.1. In general, we expect that prices must decline in order for quantity sold to increase. The points plotted in Fig. 17.1 do not exhibit the expected relationship. However, during this 10 year period, the general price level was increasing as shown by the CPI reported in Table 17.5. If we wish to deflate the observed price, we must select an appropriate index. If we planned to evaluate broiler prices within the framework of overall consumption, it would be appropriate to use the index for all items. However, in this simple analysis, it seems more appropriate to use the index for food.

TABLE 17.5. BROILER PRICE (OBSERVED AND DEFLATED) AND PER CAPITA QUANTITY—1966 - 1975

Year	CPI All Items	Food	Observed Broiler Price	Deflated Broiler Price	Per Capita Consumption
	——1967 = 100——		¢/lb	¢/lb	lb/Year
1966	97.2	99.1	41.3	41.7	32.2
1967	100.0	100.0	38.1	38.1	32.7
1968	104.2	103.6	39.8	38.4	33.0
1969	109.8	108.9	42.2	38.8	35.1
1970	116.3	114.9	40.8	35.5	36.9
1971	121.3	118.4	41.0	34.6	36.7
1972	125.3	123.5	41.4	33.5	38.4
1973	133.1	141.4	59.6	42.1	37.4
1974	147.7	161.7	56.0	34.6	37.5
1975	161.2	175.4	63.3	36.1	37.2

In order to convert nominal prices into real prices, we divide each observed price by the corresponding index number (converted from a percentage to a decimal fraction):

$$1966$$
$$41.7 = 41.3 \div 0.991$$
$$1967$$
$$38.1 = 38.1 \div 1.000$$
$$1968$$
$$38.4 = 39.8 \div 1.036$$

Figure 17.2 shows deflated broiler prices versus per capita consumption. With the exception of the observation for 1973, these data are very well behaved. The expected inverse relationship between price and quantity is clearly illustrated.

By using the Consumer Price Index, the Wholesale Price Index (WPI), or other indices published in the Survey of Current Business or other government publications, it is possible to reduce the distortion caused by inflation before beginning an analysis of a statistical series.

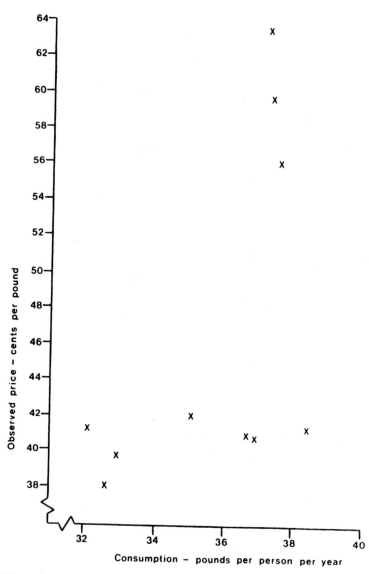

FIG. 17.1.   OBSERVED BROILER PRICE AND QUANTITY FROM TABLE 17.5

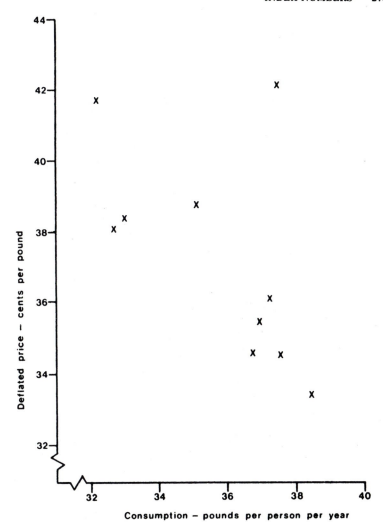

FIG. 17.2. DEFLATED BROILER PRICE AND QUANTITY FROM TABLE 17.5

Agricultural Parity

The absolute level of prices is not as important as the relative level of prices. This concept was embodied in the Agricultural Adjustment Act of 1933 which defined the concept of agricultural parity. The agricultural parity ratio is based on two indices: the index of prices received by farmers and the index of prices paid by farmers. The purpose of these calculations is to determine whether or not agricultural commodities command (or can be exchanged for) the same bundle of goods and services currently that they commanded in the base period (i.e., 1910–1914).

Prices Received by Farmers.—The Index of Prices Received by Farmers is published monthly by the U.S. Dept. of Agriculture in *Agricultural Prices*. This index is calculated using the Laspeyres Formula. The weights are based on the marketings of 55 commodities during 1953–1957 that represented approximately 93% of the total value of marketings during that period. The index is published for the base 1910–1914 = 100 as required by law and for the base 1967 = 100 for comparison with other published indices that use the 1967 base.

Prices Paid by Farmers.—The Index of Prices Paid by Farmers for Commodities and Services, Including Interest, Taxes and Farm Wages is published monthly by the U.S. Dept. of Agriculture in *Agricultural Prices*. This index is calculated using the Laspeyres Formula. More than 400 separate items used in farm production and in the farm household are included in the index. The index is published for the base 1910–1914 = 100 as required by law and for the base 1967 = 100.

Parity Ratio.—In its simplest form, the agricultural parity ratio is the ratio of the index of prices received by farmers divided by the index of prices paid by farmers. In this way, it attempts to monitor the level of agricultural outputs relative to the level of agricultural inputs including household expenses. Since changes in technology are not reflected, the ratio is essentially meaningless. Fortunately, recent agricultural legislation is moving away for the traditional use of parity prices and the agricultural parity ratio.

Dow Jones Average

The Dow Jones Average comprises 4 averages: (1) an industrial average based on the stocks of 30 manufacturing companies, (2) a transportation average based on the stocks of 20 railroads and airlines, (3) a utility average based on the stocks of 15 utilities, and (4) a composite index based on all 65 stocks. The most commonly reported of these is the Dow Jones Industrial Average (DJIA).

The DJIA is not an index in the same sense that we have examined other indices in this chapter. There is no base period comparable to those used by the Consumer Price Index or other indices. In addition, neither the Laspeyres nor Paasche formula is used. Instead, the Dow Jones Industrial average is the unweighted sum of the 30 stocks divided by an artificial divisor which is adjusted from time to time to reflect stock splits and changes in the 30 stocks included in the average. For example, on June 29, 1979, Chrysler Corp. and Esmark, Inc., were removed from the list of 30 industrials and replaced by International Business Machines Corp. and Merck & Co. The current divisor is approximately 1.44.

Since it possesses an arbitrary scale without a base period, the only real use of the DJIA is to monitor the aggregate movement over time of the prices of stocks traded on the New York Stock Exchange. The Dow Average has been criticized for being unrepresentative (it consists primarily of Blue Chip stocks) and highly volatile.

Other stock market indices which are more representative of the prices on the New York Stock Exchange are Standard & Poor's 500 stock index and the NYSE index based on all of the more than 1500 stocks traded on the Exchange.

SUMMARY

The purpose of an index is to construct a single, summary statistic which is representative of a host of statistical series. Although most indices are calculated using Laspeyres formula, some indices are constructed with a purely arbitrary scale. It is always useful to determine how an index is calculated, what weights are included, and the specific base period before attempting to use an index as an aid in decision making.

PROBLEMS

17.1. Using the following quarterly Consumer Price Index values, deflate the retail prices for broilers reported in Table 13.14 and recalculate the regression.

	Consumer Price Index		
Quarter	1975	1976	1977
I	157.0	167.1	176.9
II	159.5	169.2	180.7
III	162.9	171.9	183.3
IV	165.5	173.8	185.3

REFERENCES

For more information on the construction and use of index numbers, see:

BANNERJEE, K.S. 1975. Cost of Living Index Numbers. Marcel Dekker, New York.

FREUND, J.E. and WILLIAMS, F.J. 1958. Modern Business Statistics. Prentice-Hall, Englewood Cliffs, N.J.

MORGENSTERN, O. 1963. On the Accuracy of Economic Observations. Princeton Univ. Press, Princeton, N.J.

STOCKTON, J.R. and CLARK, C.T. 1971. Business and Economic Statistics, 4th Edition. South-Western Publishing Co., Cincinnati.

TOMEK, W.G. and ROBINSON, K.L. 1972. Agricultural Product Prices. Cornell Univ. Press, Ithaca, N.Y.

For more information on the calculation of agricultural indices, see:

U.S. DEP. AGRIC. 1975. Scope and methods of the statistical reporting service. U.S. Dep. Agric., ESCS, Misc. Publ. *1308.*

For more information on the calculation of the Consumer Price Index, see:

U.S. DEP. LABOR. 1976. Handbook of methods. U.S. Dep. Labor, Bur. Labor Stat., Bull. *1910.*

For more information on stock market indices, see:

ENGEL, L. and WYCKOFF, P. 1977. How to Buy Stocks, 6th Edition. Bantam Books, New York.

ROSENBERG, C.N. 1969. Stock Market Primer. The World Publishing Co., Cleveland.

For current releases of agricultural and business indicators, see:

U.S. DEP. AGRIC. Agricultural Prices. Crop Reporting Board, U.S. Dep. Agric., ESCS, Washington, D.C. (published periodically with an annual summary)

U.S. DEP. COMMER. Survey of Current Business. U.S. Dep. Commer., Bur. Econ. Analysis, Washington, D.C. (published monthly)

18

Time Series

Many business and economic statistics are reported as time series. That is, unlike many biological data that can be developed through planned experiments, each piece of economic data is unique. There can be no replication because we are unable to go back in time, repeat the conditions, and record a new observation. In order to address such a situation, alternative statistical approaches have evolved. These procedures are usually referred to as time series analysis. This chapter presents some of the more widespread of these techniques.

SIMPLE TIME SERIES ANALYSIS

Many statistical series exhibit repeating patterns of behavior. Daily trading of commodity futures prices, spot market prices of agricultural products, sales for a firm or industry, the quoted price per share of a traded stock or group of stocks, etc., often exhibit surprising regularity when plotted over time. The simplest approach to examining such data assumes that each observation comprises 4 components: (1) trend, (2) seasonal effect, (3) cyclical effect, and (4) an irregular effect or error term.

Although some authors describe the model as additive (i.e., $Y = T + S + C + I$), it is more common to work with the multiplicative model which is illustrated in this chapter.

$$Y = T \times S \times C \times I$$

where Y = sales, price or other variable under analysis
T = the trend component
S = the seasonal component
C = the cyclical component
I = the irregular component or error term

Figure 18.1 illustrates the basic concept of each of these 4 components. Trend represents the broad long-term growth or decline of the industry or firm. The seasonal component represents annually recurring forces that

277

affect sales or prices. The cyclical component is a measure of forces that act as broad irregular waves. Such cycles may be due to demographic changes, general business cycles, etc. The irregular component is simply an error term and represents variations that can not be attributed to trend, season, or cycle.

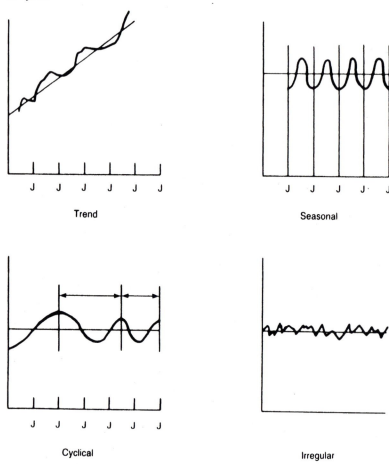

FIG. 18.1.   THE 4 BASIC COMPONENTS OF TIME SERIES ANALYSIS

Figure 18.2 shows 3 years of monthly sales data from Table 18.1 plotted over time. There appears to be both a general upward trend and a recurring seasonal pattern. The series is too short to exhibit any cyclical behavior. Trend and seasonal components can be calculated directly from a time series. Cyclical patterns are observed indirectly after the data have been adjusted for trend and seasonal components.

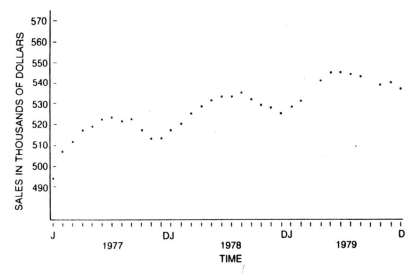

FIG. 18.2.   SALES DATA BY MONTHS FROM TABLE 18.1

TABLE 18.1.   MONTHLY SALES DATA IN THOUSANDS OF DOLLARS

Month	1977	1978	1979
January	494	517	528
February	507	520	531
March	512	525	532
April	517	528	541
May	519	531	545
June	522	533	545
July	523	533	544
August	521	535	543
September	522	532	540
October	517	529	539
November	513	528	540
December	513	520	537

Trend

Trend is usually computed in the form of a linear regression where the only explanatory variable is time. In our case

$$Y = a + b X$$

where   Y = 494, 507, 512, . . ., 537
         X = 1, 2, 3, . . ., 36

Using the procedures shown in Chapter 12, our trend line would be:

$$Y = 509.14 + 0.979 X$$

However, in more complicated situations the trend line could be estimated as a curvilinear function by using a logarithmic transformation or adding a quadratic term:

$$\log Y = \log a + b \log X$$

or

$$Y = a + b_1 X + b_2 X^2$$

Regardless of the specific regression form selected, the purpose of estimating a trend line is to extract from the data the changes that are caused by broad secular forces.

Season

The seasonal component is estimated independently using a centered 12-month moving average. This calculation yields what is commonly called the specific seasonal (SS) which is composed of both the seasonal component and the error term.

$$Y = T \times S \times C \times I$$

$$SS = \frac{Y}{\text{moving average}} = \frac{T \times S \times C \times I}{T \times C} = S \times I$$

After specific seasonals have been calculated for a series, we can average the value for each month and thereby construct a seasonal index which is free from the effects of trends and cycles.

Table 18.2 shows the worksheet and the computations necessary to estimate the specific seasonals. Although the computations are somewhat burdensome, they can readily be performed by a clerk or programmed for a computer.

Cycle

After the trend and seasonal components have been estimated, the data can be transformed and examined for evidence of cycles. This last step is highly subjective and should be approached with extreme caution.

The procedure is as follows:

TABLE 18.2. WORKSHEET FOR CALCULATING A SEASONAL COMPONENT

(I) Year and Month		(II) Sales	(III) 12-Month Moving Total Centered	(IV) 2-Year Moving Total	(V) 12-Month Moving Average Centered (Col. IV/24)	(VI) Specific Seasonal (Col. II/Col. V)
1977	Jan.	494				
	Feb.	507				
	Mar.	512				
	Apr.	517				
	May	519				
	June	522				
			6180			
	July	523		12,383	516.0	101.4
			6203			
	Aug.	521		12,419	517.5	100.7
			6216			
	Sept.	522		12,445	518.5	100.7
			6229			
	Oct.	517		12,469	519.5	99.5
			6240			
	Nov.	513		12,492	520.5	98.6
			6252			
	Dec.	513		12,515	521.5	98.4
			6263			
1978	Jan.	517		12,536	522.3	99.0
			6273			
	Feb.	520		12,560	523.3	99.4
			6287			
	Mar.	525		12,584	524.3	100.1
			6297			
	Apr.	528		12,606	525.3	100.5
			6309			
	May	531		12,633	526.4	100.9
			6324			
	June	533		12,660	527.5	101.0
			6366			
	July	533		12,683	528.5	100.9
			6347			
	Aug.	535				
	Sept.	532				
	Oct.	529				
	Nov.	528				
	Dec.	525				
1979	Jan.	528				
	etc.					

(1) The series is adjusted for trend:

$$Y' = \frac{Y}{T} = \frac{T \times S \times C \times I}{T} = S \times C \times I$$

That is, a new observation (Y') is defined by dividing each observation by its estimated trend value.

(2) The adjusted series (Y') is further adjusted for seasonal effects:

$$Y^* = \frac{Y'}{S} = \frac{S \times C \times I}{S} = C \times I$$

That is, a seasonally and trend adjusted series (Y*) is defined which is composed of only cyclical and irregular movements.

(3) The final series (Y*) is plotted over time to determine whether or not any regularity exists.

By definition, cycles last longer than 1 year and less than the entire length of the series. With such a loose definition, there is no generally accepted method for estimating a cyclical component.

SEASONAL ADJUSTMENT

One of the more important uses of time series analysis is to construct a seasonal index. Many products display distinct seasonal patterns in sales and/or prices. If sales increase from May to June, is this an indication of increased business activity or merely a reflection of the normal seasonal change? By estimating the seasonal index, a firm can decide whether month to month (or quarter to quarter) changes reflect true changing business conditions or merely a recurrence of normal seasonal movements.

In a similar way, index numbers can be adjusted for seasonality. For example, the Consumer Price Index (CPI) is reported in both a "Seasonally Adjusted" and a "Not Seasonally Adjusted" series. Table 18.3 illustrates the process of calculating a seasonally adjusted index. The first column represents the unadjusted index calculated with a Laspeyres Formula (see Chapter 17) for the base 1967 = 100. The second column reports a seasonal index calculated from the original data. Since an average month equals 100, the sum of the index values for 12 months equals 1200. These values repeat

TABLE 18.3.   CALCULATION OF A SEASONALLY ADJUSTED PRICE INDEX

Year and Month		Price Index 1967 = 100	Seasonal Index Average Month = 100	Seasonally Adjusted Index 1967 = 100
1978	Jan.	199.5	109	183.0
	Feb.	200.1	105	190.6
	Mar.	201.2	99	203.2
	Apr.	203.5	93	218.8
	May	200.3	90	222.6
	June	199.8	85	235.1
	July	200.7	90	223.0
	Aug.	202.4	96	210.8
	Sept.	205.6	101	203.6
	Oct.	217.8	110	198.0
	Nov.	220.3	112	196.7
	Dec.	223.9	110	203.5
1979	Jan.	223.2	109	204.8
	Feb.	225.4	105	214.7
	Mar.	228.5	99	230.8
	Apr.	229.0	93	246.2
	May	227.5	90	252.8
	June	226.3	85	266.2

annually until we decide to update our seasonal index and calculate new monthly values.

The third column reports the seasonally adjusted index which is calculated as follows:

$$
\begin{array}{llll}
1978 & \text{Jan.} & 183.0 = 199.5 \div 1.09 \\
& \text{Feb.} & 190.6 = 200.1 \div 1.05 \\
& \text{Mar.} & 203.2 = 201.2 \div 0.99 \\
& \text{etc.}
\end{array}
$$

In this way, we see that although the unadjusted index seems to indicate that prices are rising slowly during the first few months of 1978, they are actually rising quite rapidly. The seasonal index indicates that "on average" prices should decline during this period. Consequently, the Seasonally Adjusted Index shows a rapid advance in prices.

SUMMARY AND A WORD OF CAUTION

Time series analysis is important because it provides a technique for working with business statistics which occur as unique points in time. However, the simple structure used to decompose observed variation into the 4 components of trend, seasonal, cyclical, and error should not be mistaken for an analysis of the observed system. It is merely a statistical technique that can be useful as a first approximation or when nothing else is available.

For example, seasonal factors may be due to shifts in supply and demand which could be measured directly. In a similar manner, trend may be the result of population changes. As a result, examining the data in per capita terms may be far more revealing than studying totals.

In short, plotting the points over time can often provide a useful insight into what has occurred. In addition, the procedures discussed in this chapter may help focus attention on whether long-term trends or seasonal factors are more important in terms of future events. Armed with this information, the decision maker can concentrate on trying to understand and estimate the forces that give rise to seasonal or trend shifts.

PROBLEMS

18.1. Elsie's Elegant Eggs exhibited the following monthly pattern of prices (cents/dozen) for cartoned Grade A eggs:

		Year	
Month	1976	1977	1978
Jan.	73.2	79.0	57.9
Feb.	67.4	76.6	64.0
Mar.	62.4	68.0	64.3

Month	1976	Year 1977	1978
Apr.	62.0	63.3	59.0
May	63.3	56.1	54.6
June	62.3	57.5	51.3
July	67.4	63.1	61.8
Aug.	72.2	62.7	63.4
Sept.	63.5	62.5	65.0
Oct.	71.4	56.3	62.3
Nov.	76.7	58.4	68.8
Dec.	82.5	63.3	73.3

(a) Plot these prices over time.
(b) Calculate the secular trend and construct a seasonal index.
(c) Find the monthly Consumer Price Index, as published in the Survey of Current Business, for this period, deflate by the CPI for all items and redo steps (a) and (b).

18.2. Compute the specific seasonal of the following data:

Year and Month		Sales
1978	Jan.	725
	Feb.	738
	Mar.	740
	Apr.	745
	May	742
	June	748
	July	754
	Aug.	756
	Sept.	764
	Oct.	766
	Nov.	769
	Dec.	772
1979	Jan.	746
	Feb.	751
	Mar.	753
	Apr.	758
	May	756
	June	761
	July	764
	Aug.	765
	Sept.	769
	Oct.	769
	Nov.	773
	Dec.	776

REFERENCES

For further discussion on time series analysis and the methods of calculation, see:

CLARK, C.T. and SCHKADE, L.L. 1969. Statistical Methods for Business Decisions. South-Western Publishing Co., Cincinnati.

FREUND, J.E. and WILLIAMS, F.J. 1958. Modern Business Statistics. Prentice-Hall, Englewood Cliffs, N.J.

HARNETT, D.L. 1970. Introduction to Statistical Methods. Addison-Wesley Publishing Co., Reading, Mass.

LAPIN, L.L. 1978. Statistics for Modern Business Decisions, 2nd Edition. Harcourt Brace Jovanovich, New York.

NETER, J. and WASSERMAN, W. 1966. Fundamental Statistics for Business and Economics, 3rd Edition. Allyn and Bacon, Boston.

SPURR, W.A. and BONINI, C.P. 1967. Statistical Analysis for Business Decisions. Richard D. Irwin, Homewood, Ill.

STOCKTON, J.R. and CLARK, C.T. 1971. Introduction to Business and Economic Statistics, 4th Edition. South-Western Publishing, Co., Cincinnati.

For a discussion of statistical techniques for attempting to quantify the underlying economic forces, see:

FOOTE, R.J. 1955. A comparison of single and simultaneous equation techniques. J. Farm Econ. 37, 975−990.

FOOTE, R.J. and FOX, K.A. 1954. Analytical tools for measuring demand. U.S. Dep. Agric. Agric. Handb. 64.

KMENTA, J. 1971. Elements of Econometrics. Macmillan Publishing Co., New York.

SCHULTZ, H. 1938. The Theory and Measurement of Demand. Univ. of Chicago Press, Chicago.

TOMEK, W.G. and ROBINSON, K.L. 1972. Agricultural Product Prices, Cornell Univ. Press, Ithaca, N.Y.

WAUGH, F.V. 1964. Demand and price analysis: Some examples from agriculture. U.S. Dep. Agric. Tech. Bull. 1316.

WAUGH, F.V. 1966. Graphic analysis: Applications in agricultural economics. U.S. Dep. Agric. Agric. Handb. 326.

WOLD, H. and JUREEN, L. 1953. Demand Analysis. John Wiley & Sons, New York.

19

Control Charts

A major contribution toward using statistics to aid management was the development of the control chart (Shewhart 1931). Although control charts may be used in a "cookbook" fashion without understanding the underlying statistical foundation, statistical knowledge greatly enhances the value of control charts.

In its simplest form, a control chart provides a mechanism for monitoring performance of a production stream calling for action only when there is *statistical* evidence that the process is deviating from the established norm.

Indication for action provided by the control chart is based on statistical probability, usually better than 0.99, so that there is good assurance that action when called for is really needed, at the same time preventing changes when in fact they are not needed. Perhaps of greater importance than this statistical assurance is the psychological bonus of providing workers with definite information as to what they are expected to do, and immediate and continuous recognition of a job well done.

Control charts may be classified as variables or attributes types. Variables charts are reserved primarily for the important factors of quality which can be reported separately, while attributes charts are used when different types of quality attributes can be grouped together and reported jointly. Variables charts are based on the normal distribution, while attributes charts are based on the binomial distribution, of which the Poisson distribution is a close approximation.

The control chart may be thought of as a frequency distribution curve but placed on its side, that is, with the quality measurements on the vertical scale, and the frequency extended on a time series (Fig. 19.1).

THE VARIABLES CONTROL CHART

The Shewhart control chart for variables, or the $\bar{X}$, R chart, is undoubtedly the most generally used statistical tool in any quality control program. Since it is a variables criterion, it is naturally reserved for use with inspection data that are obtained in the form of actual measurements on some numerical scale. Since construction and maintenance of such charts in-

286

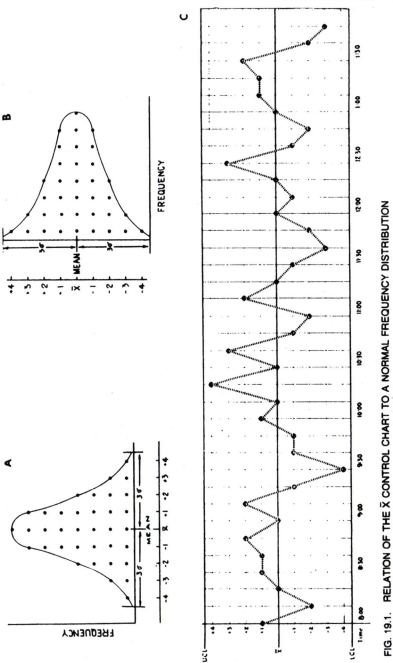

FIG. 19.1.   RELATION OF THE X̄ CONTROL CHART TO A NORMAL FREQUENCY DISTRIBUTION

A—Frequency distribution with frequency scale vertical.
B—Frequency distribution with frequency scale horizontal.
C—Control chart; frequency distribution extended into a time series.

volves a recognizable amount of time and effort, they should not be used indiscriminately, but only where it can be definitely shown that their use improves the overall operation. Since one control chart can be used for only one quality attribute, the attributes for which the charts are used should be selected with care.

The most common, and usually the first application of variables charts in food manufacture is in weight control.

It is noteworthy that control charts have been found to be extremely effective when filling is done either by machine or by hand. In fact, a worker educational program based on control charts is potentially of greater worth than the control of mechanical fillers. Even where 100% weighing is done, the control chart can be very effective. Obviously the value in such instances is largely psychological, in that an out of control indicator is an alarm to the weighers that they are not watching the scales. Dramatic savings in material and improvements in uniformity can frequently be demonstrated in a very short time. Hence this specific application has been used to "sell" a quality control program. At times such selling is overdone, so that before the dust settles, control charts blossom out everywhere, whether they are needed or not.

On the other hand, $\bar{X}$, R charts should not be limited to weight control, since their fields of application are very wide. Some other areas in which these charts may be, and are being, used in the food industry are: volume of containers, sizes or dimensions, color and other appearance properties; consistency, firmness, and other rheological properties; moisture, fat, protein, and other chemical or nutritional properties; mold, bacterial, and other microanalytical counts or measurements; yields, batch or continuous mixing operations, etc.

Although such control charts are usually associated with the production line, they may also be maintained for charting quality of incoming materials. Thus, for example, a control chart may be maintained for all the raw material entering the plant, and a second for the same material after sorting. In addition to the use of such a pair of charts in maintaining control of quality, they may also serve as a basis for incentive pay to the workers on the sorting belt, and as basic data for cost accounting. Other personnel, suppliers, or sales performances by individuals or groups can also be charted in this manner. In fact, any type of operation for which numerical evaluations are available can be charted, providing such charting does more than pay for itself.

Preliminary Considerations

Certain decisions must be made, essentially on the what, how, where, and when of the problem before constructing the control chart.

What Is to Be Measured?—At times this is perfectly obvious, as in the case of weight control of fluid materials. At other times it is more difficult to determine just what to measure, as in the case of canned peaches which may contain from 5 to 9 halves. If, for example, the fifth half just fulfills the required weight, then the pack is filled economically. However, if the fifth peach just misses the required fill, the can must be overfilled by 20% to meet the minimum fill requirements. Obviously charting the fill alone will not provide the needed control. What is needed here is size control so that a given number of peach halves of a certain size will just meet the required minimum fill.

How Is It to Be Measured?—The method or procedure of obtaining the measurement should be objective, precise, and accurate. The procedure should be objective, in that it should be instrumental, rather than a matter of personal opinion. It should be accurate in that it actually measures the quality attribute it is purported to measure and it should be sufficiently precise to take advantage of the process capability. For example, if a filler can be adjusted to discharge quantities within a tolerance of ±1.0 g, it would be folly for the inspector to use a balance capable of weighing only to the nearest ounce.

Where Is It to Be Measured?—Ordinarily the best location for maintaining the control station and performing the necessary tests is right at the point of operation, rather than in the laboratory. If fill of container is to be checked, it is advantageous to do the weighing and the charting right at the filler. It may be difficult to perform some test procedures right on the production line, as, for example, some chemical determinations which must be done in a laboratory. The control charts, nevertheless, should be posted at the point where those responsible for the immediate operation may see a continuous record of their performance and take the necessary action immediately.

When Is It to Be Measured?—The frequency of measurement may be determined in a manner similar to the determination of sample size, particularly if the purpose of the measurement is to establish the grade, or quality level, of lots for shipment. For example, if the sampling plan calls for the inspection of 500 units in lots of 50,000, approximately every 100th unit could be inspected. However, if the main purpose of the inspection is to detect an out of control situation as soon as possible so that corrective action can be taken, then the major criterion is some knowledge of how rapidly, or when, an out of control situation is likely to develop, and the cost of such an out of control situation, for each unit of time, as compared to the cost of inspection. For example, if it is generally known that loads of raw material delivered to the plant are essentially the same in quality within each load, but may differ substantially from load to load, it would be logical to check the product thoroughly as soon as possible after a new load of material is used. If loads of tomatoes enter a plant and are utilized in consecutive order, it would be advisable to time the mold determination to each new lot, in order to determine whether enough tomatoes are sorted out, or too many.

Ordinarily a smoothly running operation requires that operations be performed at regular time intervals. This is desirable for quality control inspections, too, but the time intervals should be varied to a sufficient extent to avoid anticipation of the moment of inspection by the operators.

How Many Observations at One Time?—The ordinary $\bar{X}$, R chart requires that a number of observations be made, otherwise there is no opportunity to obtain an average ($\bar{X}$), or a range (R). In some rare instances individual observations may be charted directly. This might be the case where the observation involves a long and expensive procedure. Ordinarily, 4 to as many as 10 observations are taken at 1 time interval, with 5 being the most usual, perhaps because the mean can be so easily determined from the sum of the 5 observations (multiply by 2 and divide by 10). At times, a logical number of observations becomes obvious, as in the case of a number of molds, in the line. The number of observations at one time can consist of 1 per mold, thus providing an opportunity for the range to cover all molds at every observation period.

CONSTRUCTING THE VARIABLES CONTROL CHART

After first considering the questions of how, where, and when it is to be used, the control chart may be constructed and adjusted in accordance with the following three steps.

(1) Determine Process Capability.   This is a preliminary study to determine just how well the uniformity of the particular quality characteristic can be maintained under the current operating conditions.

(2) Adjustments to Meet Specifications at Minimum Costs.   Assuming the variability in quality is a characteristic of the process, should changes be made in the mean quality level, to meet specifications or to reduce costs?

(3) Adjustments to Improve Performance.   Can the control chart be used to indicate means of reducing variability in the quality level thereby resulting in even greater product uniformity and savings?

Determining Process Capability.—In constructing this tentative control chart, we are essentially assuming that the average variability that we find in a group of observations taken at one time period should be the same at all times, so that if we maintain the same average quality level over long periods of time, we should be in control. The $\bar{X}$ chart thus provides us with the information as to whether the average, or mean value has drifted, or is still in control. Similarly the R, or range chart, indicates whether the inherent variability in the process has changed.

Our working tools are the factors in Table 19.1 which are short-cut multipliers for determining acceptance-rejection limits. We proceed to obtain a number of samples at different time intervals. Common practice is to take 10 groups of 5 observations for a total of 50, and the tentative control limits are established on the basis of these 50 determinations (Shainin 1950).

Example 19.1.   Setting Up Preliminary Control Limits.—We wish to construct a chart for controlling fill of a 12 pocket FMC filler while filling frozen cut wax beans, in terms of tenths of ounces (2.8 g) above or below the 10 oz (283.5 g) minimum fill. While the filler is operating routinely, we

TABLE 19.1.   FACTORS FOR COMPUTING 3-SIGMA CONTROL CHART LIMITS FOR $\bar{X}$, R CHARTS

No. of Measurements in Sample n	Chart for Averages Factors for Control Limits A_2	Chart for Ranges Factors for Control Limits D_3	D_4	Factors for Standard Deviation d_2
2	1.880	0	3.268	1.128
3	1.023	0	2.574	1.693
4	0.729	0	2.282	2.059
5	0.577	0	2.114	2.326
6	0.483	0	2.004	2.534
7	0.419	0.076	1.924	2.704
8	0.373	0.136	1.864	2.847
9	0.337	0.184	1.816	2.970
10	0.308	0.223	1.777	3.078
11	0.285	0.256	1.744	3.173
12	0.266	0.284	1.717	3.258
13	0.249	0.308	1.692	3.336
14	0.235	0.329	1.671	3.407
15	0.223	0.348	1.652	3.472

Formulas

For control limits of:

Averages – U.C.L. $= X + A_2\bar{R}$

L.C.L. $= X - A_2\bar{R}$

Range – U.C.L. $= D_4\bar{R}$

L.C.L. $= D_3\bar{R}$

Estimate of Standard Deviation or Sigma

Sigma $= \bar{R}/d_2$

Source: Reproduced from Kramer and Twigg (1970).

collect 5 consecutive packages, weigh them to the nearest tenth oz (g), and return them to the line. We repeat this 10 times, and tabulate the results as shown in Table 19.2. We then proceed to obtain a sum (Σ) for each column, that is, each group of 5. Dividing this sum by 5, we obtain a mean ($\bar{X}$) for each group of 5. The range (R) is obtained by subtracting the smallest value in each column from the largest value. Thus for example in the first group shown in Table 19.2, the highest value for group 1 is 4, and the lowest value is -2. Thus R $= 4 - (-2) = 6$. Totaling all 10 means, and dividing by 10, we obtain the mean of means, or $\bar{X} = 16.6/10 = 1.66$. This value is the tentative central line for the $\bar{X}$ chart. The mean range ($\bar{R}$) is obtained by summing the individual ranges for the 10 groups, dividing by 10, or $\bar{R} = 40/10 = 4$. This value is the tentative central line for the R chart.

To obtain the upper control limits for the $\bar{X}$ chart, we multiply the mean range by the appropriate A_2 value from Table 19.1 and add the product to the mean. Since there are 5 observations per group, or n $= 5$, we select the

TABLE 19.2.   PRELIMINARY DATA FOR CONSTRUCTING $\bar{X}$ AND R CONTROL CHARTS FOR FILL OF CONTAINER

Within Group	Groups									
	1	2	3	4	5	6	7	8	9	10
a	2	4	3	0	2	1	3	4	2	0
b	0	3	2	2	0	-2	4	6	4	2
c	4	4	4	-1	3	0	2	2	0	-2
d	1	0	2	2	1	2	1	3	2	3
e	-2	3	2	-2	0	0	2	1	2	2
Sum (Σ)	5	14	13	1	6	1	12	16	10	5
Mean ($\bar{X}$)	1.0	2.8	2.6	0.2	1.2	0.2	2.4	3.2	2.0	1.0
Range (R)	6	4	2	4	3	4	3	5	4	5

$\Sigma \bar{X} = 16.6$

$\Sigma R = 40$

$\bar{\bar{X}} = 1.66$

$\bar{R} = 4$

$\text{U.C.L.}_{X} = \bar{\bar{X}} + A_2 R = 1.66 + (0.577)(4) = 3.97$

$\text{L.C.L.}_{X} = \bar{\bar{X}} - A_2 R = 1.66 - (0.577)(4) = -0.65$

$\text{U.C.L.}_{R} = D_4 R = (2.114)(4) = 8.46$

$\text{L.C.L.}_{R} = D_3 R = (0)(4) = 0$

A_2 value of 0.577 from Table 19.1, and the upper control limit (U.C.L.) for the mean chart is:

$$\text{U.C.L.}_{X} = \bar{\bar{X}} + A_2 R = 1.66 + (0.577)(4) = 3.97$$

The lower control limit (L.C.L.) is obtained similarly, except that the product of $A_2 R$ is subtracted from the mean:

$$\text{L.C.L.}_{X} = \bar{\bar{X}} - A_2 R = 1.66 - (0.577)(4) = -0.65$$

The upper control limit (U.C.L.) for the range chart is determined by multiplying the mean range by the appropriate D_4 value from Table 19.1:

$$\text{U.C.L.}_{R} = D_4 R = (2.114)(4) = 8.46$$

The lower control limit (L.C.L.) for the range chart is determined by multiplying the mean range by the appropriate D_3 value from Table 19.1. Since D_3 for observations up to 6 (n = 6) is 0, the L.C.L. = $D_3 R$ = (0) (4) = 0.

We now have all the necessary data to set up our tentative variables control chart, as shown in Fig. 19.2.

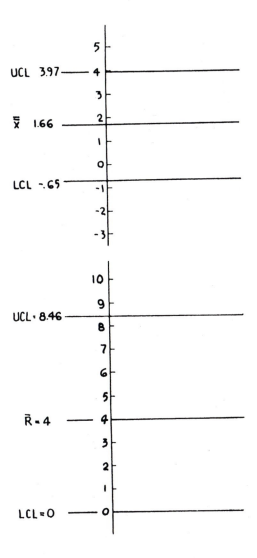

FIG. 19.2. TENTATIVE CONTROL CHART FOR VARIABLES, BASED ON DATA SHOWN IN TABLE 19.2

It may be of interest to discuss the meaning of the short-cut multipliers in Table 19.1, particularly A_2. The standard control chart provides the assurance that over 99% of the items in an operation will be included within the control limits, provided the operation is in control. Since this 99%+ assurance may be obtained only if we use limits that are about 3 standard deviations from the mean, control charts are set up so that the upper and lower control limits are, respectively, 3 sigmas (standard deviations) above and below the mean. We also know that average values are more precise than single values, so that as the sample number is increased, precision is increased by the reciprocal of the square root of the number in the sample. We could therefore calculate the upper and lower control limits by transforming the data into a standard normal deviate: control limits = $\bar{X}$ ± $ks/\sqrt{n}$ where the number of sigmas, $k = 3$, since we are interested in 99%+ assurance. The estimate of the standard deviation is s, which is obtainable by dividing the range by the appropriate d_2 value, which in this instance is 2.326 as found in Table 19.1. Thus $s = R/d_2 = 4/2.326 = 1.72$. Of course, n is 5. Thus control limits for the mean chart $= ± (3)(1.72)/\sqrt{5} = 2.31$, which is the same as $(0.577)(4) = 2.31$. Since $ks/\sqrt{n} = A_2R$, $s = R/d_2$, and k is always 3, thus

$$\frac{3R/d_2}{\sqrt{n}} = A_2R, \text{ or } A_2 = \frac{3/d_2}{\sqrt{n}}$$

In our example

$$A_2 = \frac{3}{2.326} \Big/ 2.236 = 0.577$$

the value found in Table 19.1 for A_2 with $n = 5$.

Adjustments to Meet Specifications.—In the preceding pages we have constructed control charts on the basis of current performance. We now wish to compare this performance to specifications in order to determine whether current performance is capable of meeting the specifications. It is also possible that current performance is exceeding specifications, so that savings could be made in materials or labor.

This calculation can be made on the basis of the standard deviation(s) which we have already calculated in Example 19.1 as $R/d_2 = 4/2.326 = 1.72$. In stating our problem, we indicated that the values shown in Table 19.2 are in tenths of ounces (2.8 g) in excess of 10 oz (283.5 g), which is the minimum fill. If we further specify that we want to achieve practically complete freedom from underfills, we should place the mean at least 3 standard deviations above the declared minimum, in order to have better than 99% assurance that no individual package will weigh less than 10.0 oz (283.5 g). Our mean line ($\bar{X}$) therefore should be at least 10 oz (283.5 g) plus 3 standard deviations, or

$$10.0 + (3)(1.72) = 10 + 5.16 \text{ tenths, or } 10.516$$

Since our mean fill is only 10.166 (Table 19.2), we are not meeting our specifications. We can in fact, by reference to Appendix Table B.1, determine how frequently we can expect a package to be underfilled. Since the average is 0.166 above mean, and standard deviation is 0.172 tenths of oz, then average is 0.166/0.172, or 0.97 sigmas above the minimum. Using

Appendix Table B.1, we find that with 0.97 sigmas we are covering 33.4% of the distribution. Added to the 50% of the distribution above the mean, we may expect 83.4% of the packages to be above the minimum. As many as 16.6% of the packages, however, may be expected to fall below the specified minimum weight.

When faced with these conclusions, that is, that more than 0.5 oz (14.2 g) of material must be given away with every package, in order to achieve practically complete freedom from underfills, management reaction may be to accept the situation, modify the specification, or initiate a study to improve the situation. Management may prefer a compromise. The specifications may be modified to state that fill should be so controlled that not more than 10% of the individual packages may contain less than 10 oz (283.5 g). This means that we are looking for that value of k (Appendix Table B.1), the multiplier for sigma which will cover 40% or 0.40 of the area under the normal distribution curve. This 40%, when added to the 50% of the distribution above the mean, will cover 90% of all the packages. Using Appendix Table B.1, we find this value of k is 1.29 standard deviations. Thus, the mean should be set for the minimum weight +

$$\frac{ks}{10(\text{tenths oz})} \quad \text{or} \quad = \quad 10.0 \quad + \quad \frac{(1.29)(1.72)}{10} \quad = \quad 10.222$$

Assuming that variability in terms of the performance of the filler is unchanged, A_2R remains 0.231 as in Example 19.1 and the new control limits are 10.22 ± 0.23, as shown in Fig. 19.3. Limits for the R charts, of course, do not change as a result of this shift in the $\bar{X}$.

Indications for Action.—The control chart fulfills two equally important functions. One is to indicate when action, that is, an adjustment in the operation, is needed. The other is to prevent adjustment when such action is not needed. Without statistical guides which the control chart provides, the operator is frequently tempted to adjust a process simply because the measurement he obtains on a sample indicates that it is somewhat below or above average. If the process is actually in control, and the chance reading obtained was within control limits, then any such change will tend to throw the operation out of control subsequently. This situation is met frequently, and can be easily recognized by excessive fluctuations which can be reduced only by full knowledge of the inherent variability of the process. Thus process adjustments are justified only under the following conditions:

(1) Point on $\bar{X}$ chart above U.C.L.—Reduce fill (Fig. 19.4A)
(2) Point on $\bar{X}$ chart below L.C.L.—Increase fill (Fig. 19.4B)
(3) Six or more consecutive points on $\bar{X}$ chart all above or all below mean line.—Recalculate limits and adjust filler accordingly. This means that limits can be tightened and further savings be made (Fig. 19.4C).
(4) Point on range chart above U.C.L.—Individual pocket trouble. Examine individual pockets, and if cause is not found, repeat check on pocket uniformity.

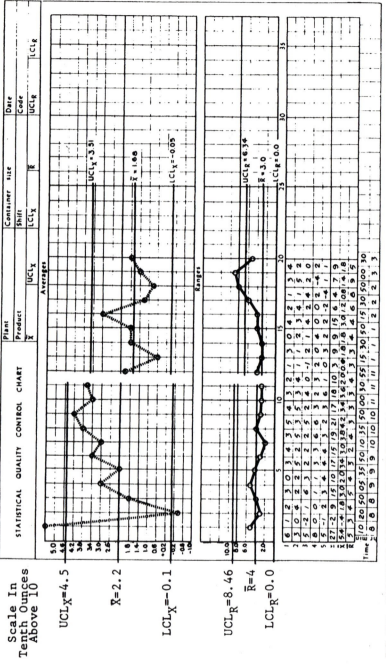

FIG. 19.3. VARIABLES CONTROL CHART FOR FILL OF CONTAINERS

1 oz = 28.4 g.

(5) Six or more consecutive points on $\bar{R}$ chart above the mean line.—Proceed as in (4) above.

(6) Six or more consecutive points on $\bar{R}$ chart below the mean line.—Proceed as in (3) above.

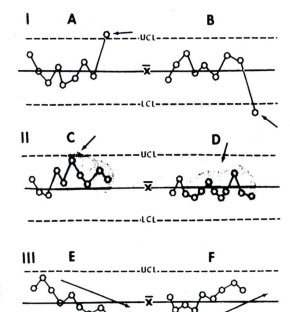

FIG. 19.4. CONTROL CHART INDICATIONS FOR ACTION

I—Beyond control limits.
II—Shifts. III—Trends.

Example 19.2. Recalculation of Control Limits.—The control chart with limits as indicated for permitting a maximum of 10% underfills is shown as Fig. 19.3. Adjustments are indicated as follows:

(1) At 8:10 (first point) average weight above U.C.L.—Fill must be reduced.

(2) At 8:20 (second point) average weight below L.C.L.—Fill must be increased.

(3) Points 3 to 5 showed normal deviations, but points 6 to 11 were all above the mean line, indicating that control limits should be recalculated, and fill reduced. Also, consistently low values on R chart indicate recalculation. Using data of samples 6 to 11, we obtain a new $\bar{R}$ of 3.0, 1.3 sigma is therefore =

$$\frac{1.3\bar{R}}{d_2} = \frac{1.3(3)}{2.326} = 1.68$$

The new $\bar{X}$ is $10 + 1.68$ (tenths), U.C.L. is

$$\bar{X} + A_2R = 1.68 + 0.577(3) = 3.51 \text{ (tenths) and}$$

$$\text{L.C.L. is } \bar{X} - A_2R = 1.68 - 0.577(3) = -0.05 \text{ (tenths)}$$

$$\text{U.C.L. for range (U.C.L.}_R) = D_4\bar{R} = 2.114(3) = 6.324 \text{ (tenths)}$$

D_4 values from Table 19.1. The new limits are shown on the right side of the chart.

(4) Point 18 is above U.C.L. on range chart, indicating pocket trouble. Data at point 19 show trouble has not as yet been corrected. At point 20, operation is back in control, after trouble was located and corrected.

Adjustments to Improve Performance.—Control charts are not static in that once they are set up, they are used solely to take action if any points appear above the upper or below the lower control limits. As shown above, a reduction in variability may be noted and taken advantage of promptly. An out-of-control situation may be noted in other ways too, which may require a separate study before the source of variation may be located, and corrected, with further improvement in uniformity and economy.

It has already been stated that some action is indicated not only if points are beyond the control limits, but also if there is a consistent number of points all above or all below the central line (see Chapter 15). Similarly, if the points seem to be close to the central line, it is advisable to recalculate the limits to take advantage of the additional improvement in uniformity (Fig. 19.4D).

At times the chart will reveal a trend, that is, a consistent increase or decrease in value with successive points. Here, too, the rule of 6 may be used. Thus if 6 points are successively higher, it is sufficient proof that a trend is present, so that action is justified even if none of the points is as yet beyond the control lines. Something of this nature can occur if a mold wears gradually, or an adjustment tends to shake loose with use, or if an operator becomes less effective with increasing fatigue, etc.

Points 18 and 19 in Fig. 19.3 point to an unusual variability among individual pockets of the filler, thus indicating that an overall adjustment of the filler to dispense more or less material is not the solution, but a study of the performance of each individual pocket is needed. Such a study is presented in the following example.

Example 19.3.   Improving Filler Performance.—To determine if there is a real difference in package weights, filled from different pockets of a 12 pocket filler, 1 pocket is marked as No. 1. Beginning with a package that was filled from the marked pocket, 60 consecutive packages were removed from the line and weighed. The weights, in terms of tenths of ounces (2.8 g) above 10, are presented in Table 19.3, arranged so that the 5 replications for each pocket appear in the same row. Each group of 12, containing 1 unit from each pocket, is ranked in order of decreasing weight, and the ranks totaled. Referring to Appendix Table I.1, any rank totals of less than 13 or more than 52 indicate significant differences among pockets. Once this has been established, it may be further concluded from the lower pair of entries in Appendix Table I.1, for 12 treatments and 5 replications, that any 1 pocket having a rank sum of less than 20 or more than 45, is significantly low and high, respectively. Thus, there is a significant difference between pockets, and pockets 3 and 4 underfill, while pocket 10 overfills.

TABLE 19.3.   DATA ON FILL-IN WEIGHTS FROM A 12 POCKET FMC FILLER

Pocket Marked	Weight of Units, in Tenths Ounces Above 10 Oz (283.5 g)										Mean	Range	Rank Total
	1–15		16–30		31–45		46–60		61–75				
	Wt.	Rank	Wt.	Rank	Wt.	Rank	Wt.	Rank	Wt.	Rank			
1	+2	4	0	7	−2	11.5	+4	1	0	6	+0.8	6	29.5
2	−1	11.5	0	7	−1	8.5	+1	6.5	0	6	−0.2	2	39.5
3	+1	6.5	−1	10	−2	11.5	−1	10	−3	11	−1.2	4	49.0[1]
4	0	9	−3	12	−1	8.5	−3	11.5	−4	12	−2.2	4	53.0[1]
5	+2	4	−2	11	−1	8.5	−3	11.5	−2	9.5	−1.2	5	44.5
6	+2	4	+1	3	−1	8.5	+1	6.5	−2	9.5	+0.2	4	31.5
7	+3	1.5	0	7	+2	4	+1	6.5	+3	2	+1.8	3	21.0
8	0	9	+1	3	+2	4	+3	2.5	+2	3.5	+1.6	3	22.0
9	−1	11.5	+1	3	+3	1.5	+2	4	+4	1	+1.6	5	21.0
10	+3	1.5	+4	1	+2	4	+1	6.5	+2	3.5	+2.4	3	16.5[1]
11	+1	6.5	0	7	+3	1.5	0	9	0	6	+0.8	3	30.0
12	0	9	0	7	0	6	+3	2.5	−1	8	+0.4	4	32.5

[1]Significantly high or low.

An examination of these rank totals also reveals a pattern, or a cycle in the fill weights, where beginning with pocket 1 there is a steady decrease in weight until pocket 4 is reached, and then there is a rise until pocket 10 is reached, after which there is another decrease to complete the cycle.

An examination of the filler showed that the bottom ring which forms the bottom of the pockets was even through the entire cycle; however, the top of the filler was warped to the extent of about 3/16 in. (0.476 cm), which accounted for the difference in fill among the pockets. If this warp can be eliminated entirely, it could result in a saving of approximately 0.25 oz (7.09 g) per carton, or about 2.5% of the fill-in weight.

Use of Analysis of Variance for Determining Control Chart Limits.—As indicated above, $\bar{X}$ chart control limits are based on unassignable variability which is considered to be an inherent part of the process, so that the operation should remain within control as long as the process average does not drift with time. Hence, if there should be no assignable variance within a time period [i.e., among the number (n) of samples taken at the same time] control limits calculated in accordance with the procedures described previously should be essentially the same as if they were calculated by variance analysis. If, however, there should be some assignable cause for a significant part of the variance within the group of samples tested at the same time period, then control limits based on the standard deviation (s) of the residual error of the analysis of variance should provide a tighter, and more appropriate control limit—thus in effect performing all 3 steps in the construction of the control chart simultaneously.

Example 19.4.   Deriving Control Limits by the Analysis of Variance.— Cookies are baked continuously on a belt on which they are placed 7 abreast (row). It is important to control the process in such a manner that moisture content of the baked cookies will average 12.3%. Moisture percentages for 6 rows of cookies are given in Table 19.4. Since $\bar{X}$ is exactly 12.3, there is no need to adjust the process as a whole. We may use the mean range ($\bar{R}$) and the appropriate A_2 value from Table 19.1 to determine the control limits in the usual manner as illustrated in Example 19.1. Thus:

$$\bar{X} \pm A_2\bar{R} = 12.3 \pm (0.419)(4.90)$$

or
$$12.3 \pm 2.05$$

TABLE 19.4.   DATA ON PERCENTAGE MOISTURE OF BAKED COOKIES

Cookies Within Row	1	2	Rows—% Moisture 3	4	5	6	Row Sum (Σr)
1	15.6	11.8	13.5	12.6	11.0	10.6	75.1
2	17.2	13.5	11.9	14.1	11.9	12.1	80.7
3	14.0	10.5	11.5	11.3	11.2	11.6	70.1
4	13.6	12.8	11.3	12.1	11.9	12.5	74.2
5	15.0	15.2	12.2	11.5	12.2	12.8	78.9
6	13.8	15.8	12.3	10.8	12.4	13.1	78.2
7	9.0	13.0	8.8	8.9	8.9	10.7	59.3
Column sum (Σc)	98.2	92.6	81.5	81.3	79.5	83.4	Σrc = 516.5
Column mean (Xc)	14.03	13.23	11.64	11.61	11.36	11.91	$\bar{X}$ = 12.30
Column range (Rc)	8.2	5.3	4.7	5.2	3.5	2.5	$\bar{R}$ = 4.90

Sum of squares =   6493.99
Correction factor = $(516.5)^2/42$ 6351.72

Corrected sum of squares =   142.27

Sources	Degrees of Freedom, df	Sum of Squares, s.s.	Mean Square, s^2	Variance Ratio, F
Total	41	142.27		
Rows	5	40.53	8.11	5.01
Within rows	6	53.13	8.86	5.47
Residual	30	48.61	1.62	

Standard deviation = $\sqrt{1.62}$ = 1.27

We would therefore be assuming that because of the random, i.e., unassignable variations in the process or product, we cannot expect to control the moisture of the cookies any better than within a range of 10.25 to 14.35% moisture.

A casual perusal of the data in Table 19.4, however, even without the benefit of a statistical analysis, should convince us that variance within rows is not due to chance alone, since the seventh cookie in each row is so frequently considerably drier than the others. We are, therefore, justified in attempting to remove the variance caused by cookies within the row, as well as the variations among rows, from our estimate of the residual standard deviation which will determine the position of the control limits. This can be accomplished by the analysis of variance as shown in the lower part of Table 19.4. Since the mean square of the residual is equal to the standard deviation squared, then the standard deviation is:

$$s = \sqrt{1.62} = 1.27$$

As shown in Example 19.1, control limits can also be calculated directly from the standard deviation. For the usual 3-sigma control limits, we would therefore multiply the value of 1 standard deviation by 3, and divide by the square root of n, which in this case is 7 since there are 7 cookies per row and the entire row would be taken at each time period. Thus:

$$\bar{X} \pm \frac{3s}{\sqrt{n}} = 12.3 \pm \frac{3(1.27)}{\sqrt{7}}$$

or
$$12.3 \pm 1.44$$

We can now state that if appropriate adjustments are made not only to ascertain that different lots of cookies remain the same, but that in addition the belt performs similarly across its entire width, then moisture can be maintained within a range of 10.86 to 13.74%, which is considerably narrower than the range of 10.25 to 14.35% as calculated by the range (A_2R) method.

ATTRIBUTES CONTROL CHARTS

In the preceding pages we have stated that variables charts may be applied where actual numerical measurements of the quality attributes are available. We have also stated that where such data are available it is wasteful not to take advantage of this information, since we can obtain more precise results with fewer samples by the use of variables procedures. However, when actual numerical values are not available, or feasible, so that all that can be determined is whether the unit tested is acceptable or unacceptable, we have attributes type data. In such cases an attributes chart is just as effective as a variables chart. Furthermore, if there is a substantial number of characteristics for which we test, and we cannot afford to maintain a separate chart for each, then the use of an attributes chart is practically a necessity. The various characteristics can be lumped together into a single attributes chart, or at most 2 or 3 charts, each covering that group of characteristics which reflects their importance, such as minor, major, and critical.

Thus attributes charts are used when measurements are not possible, or not practicable; and where several characteristics are to be considered simultaneously. In the case of food products, the application of attributes control charts is most effective for the defects category of quality, and occasionally for some other attributes of quality which are either present or absent, such as the absence of a color strip in candy, or absence of salt in a cake, or the presence of off-flavor. In each case the factor is either present or absent, and the actual quantity if present is unimportant

Preliminary considerations such as what, how, where, when, and how many tests are to be made apply to attributes charts as well as to variables charts described earlier. Additional consideration should be given to the grouping of the characteristics. One of the important advantages of the attributes chart is that it permits the grouping of a number of characteristics, or defects, into a single chart. Some characteristics, however, may be of greater importance than others. In the manufacture of gumdrops, for instance, the presence of a speck of carbon or a surface mark is of equal importance, and may be lumped together, each unit so marked counting as 1 defective; however, a broken piece of candy, or one from which the color is missing, is a much more serious defect. In such instances, one of two things can be done. The less important defects may be grouped together, and charted as minor defects, while the more important defects may be assigned to a second grouping, and charted as major defects (Kornetsky and Kramer 1957). Another possibility is to weigh each type of defect. Thus a carbon speck could count as 1 defect, while a broken piece might count as 2, and one which is not colored as 3, etc.

Kinds of Attributes Charts

All attributes charts are essentially similar, and the specific kind used depends largely on convenience for the specific purpose.

(1) p charts—Proportion, or percentage of defective items in a sample. For example, if there are 10 cull tomatoes in a sample of 200, the chart would show 0.05, or 5% defectives.

(2) np charts—These are used where it is more convenient to state the actual number of defective units in the sample. Thus, 10 cull tomatoes in the sample of 200 would be charted as 10.

(3) c charts—Here the number of defects, rather than the number of defective units, is reported. It is used where a number of defects may be found in a unit. Thus, for example, if 6 separate defects are found in a side of beef, the count of 6 is charted.

Construction of the p Chart

As with the variables chart, some past history is needed upon which to base tentative control limits. We must therefore obtain some data on the average occurrence of defective items. In Table 19.5, we have listed the total number of tomatoes examined during 20 days of inspection (column 2) and the number of defective tomatoes found during each day (column 3). Both these columns are summed, and the mean fraction defective ($\bar{p}$) is obtained by dividing the number of defective tomatoes by the total number of tomatoes examined:

$$\bar{p} = \frac{\text{number of defectives}}{\text{number examined}} = \frac{1121}{42,800} = 0.0262, \text{ or } 2.62\%$$

As stated previously, this type of attributes chart is based on the binomial distribution. Hence the standard deviation for fraction defective is: $\sigma = \sqrt{p(1 - p)/n}$. Since we ordinarily set our upper and lower control limits at 3 standard deviations, then the upper control limit for fraction defective would be:

$$\text{U.C.L.}_p = \bar{p} + 3\sqrt{p(1 - p)/n}$$

The lower control limit would be:

$$\text{L.C.L.}_p = \bar{p} - 3\sqrt{p(1 - p)/n}$$

Assuming that a sample of 200 tomatoes is inspected, the control limits would be:

$$\text{U.C.L.} = 0.0262 + 3\sqrt{\frac{0.0262(1 - 0.0262)}{200}} = 0.0601, \text{ or } 6.0\%$$

and

$$\text{L.C.L.} = 0.0262 - 3\sqrt{\frac{0.0262(1 - 0.0262)}{200}} = -0.0077, \text{ or } -0.77\%$$

Since the L.C.L. value is negative, the lower control limit is placed at zero, and the control chart is constructed as shown in Fig. 19.5.

In some instances it may not be convenient to have a standard sample size, so that n may vary. For such situations, the equation for the control limits may be set up a little differently as follows: control limits for fraction defective = $\bar{p} \pm 3\sqrt{p(1 - p)}/\sqrt{n}$. Thus $3\sqrt{p(1 - p)}$ may be calculated separately, and used as a constant, which may then be divided by the square

% DEFECTIVES

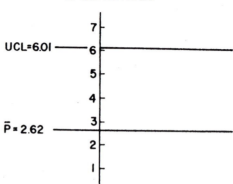

FIG. 19.5. CONTROL CHART
FOR ATTRIBUTES—p CHART

Showing the mean ($\bar{p}$) and control limits for defective tomatoes
after sorting. Limits based on
data shown in Table 19.5, n =
200.

TABLE 19.5. CALCULATIONS FOR DETERMINING MEAN FRACTION DEFECTIVE AND
CONTROL LIMITS FOR DEFECTIVE TOMATOES

1	2	3	4	5	6	7
					U.C.L.	L.C.L.
					When $\bar{p}$ = 0.0262	
	No.		Fraction		$\sqrt{p(1-p)}$ = 0.480	
	Inspected	No. of	Defective	%	and n as Shown	
Date	n	Defectives	p	Defective	in Col. 2	
August 1	1700	64	0.0376	3.76	3.78	1.46
2	2300	66	0.0286	2.86	3.63	1.62
3	2800	105	0.0375	3.75 (high)	3.53	1.71
4	2100	94	0.0447	4.47 (high)	3.67	1.57
5	600	12	0.0204	2.04	4.58	0.66
7	2200	55	0.0250	2.50	3.64	1.60
8	3500	97	0.0277	2.77	3.43	1.81
9	3300	102	0.0275	2.75	3.46	1.78
10	1200	33	0.0309	3.09	4.01	1.23
11	600	14	0.0233	2.33	4.58	0.66
12	800	6	0.0075	0.75 (low)	4.32	0.92
14	3000	38	0.0127	1.27	3.50	1.74
15	3600	85	0.0236	2.36	3.42	1.82
16	4300	92	0.0213	2.13	3.35	1.89
17	4100	112	0.0273	2.73	3.37	1.87
18	2200	60	0.0272	2.72	3.64	1.60
19	1000	31	0.0323	3.23	4.14	1.10
21	900	18	0.0200	2.00	4.22	1.02
22	1200	26	0.0217	2.17	4.01	1.23
23	1400	31	0.0221	2.21	3.90	1.34
Σ	42,800	1121				
$\bar{p}$ = 1121/42,800 = 0.0262			0.0262	2.62		

root of the number of items in the specific sample (n). Thus in our example, we would calculate $3\sqrt{p(1 - p)} = 3\sqrt{0.0262(1 - 0.0262)} = 0.480$. If n happened to be 200, as in the preceding example, the value of 0.480 would then be divided by the square root of 200, to obtain the value to be added or subtracted to the mean fraction defective ($\bar{p}$) to obtain the control limits

$$\pm 0.480/\sqrt{200} = \pm 0.0339$$

If, however, only 100 tomatoes are examined, we would use the same value of 0.480, but divide by $\sqrt{100}$. In Table 19.5 are shown limits for each individual day's operation, in which the total number of tomatoes sampled for the day were used as n. It may be of interest to note in passing that according to such calculations, the percentage of defective tomatoes on August 3 and 4 was beyond control limits, indicating poor or insufficient sorting or exceptionally high numbers of defectives in the raw material. The low number of defectives on August 12 may have been due to the small quantity of fruit being sorted by the same size labor force.

Constructing the np Chart

This is identical to the p chart, except that the actual number of defective units is recorded instead of the proportion or percentage. Thus in our example, if 200 tomatoes are used for the sample, pn for the central line would be 5.24 tomatoes instead of the $\bar{p}$ of 0.0262 or 2.62%. Similarly, U.C.L.$_{np}$ would be 12.02 tomatoes instead of 0.0601, or 6.01%, and L.C.L.$_{np}$ would be 0 as before (Fig. 19.6).

Constructing the c Chart

In contrast with the p, np, and particularly with the $\bar{X}$, R variables charts, the c chart has only limited use in the food industry. Since the c chart is ordinarily applied to relatively large units or assemblies where a count is made of the defects found in the unit, it is limited to relatively large, and expensive items. Some examples of its possible use in the food industry are in controlling the number of defects in livestock, or carcasses of large animals, cakes, prepared dinners, gift assortments.

As with the other charts, we must first develop or find some past data upon which to establish a mean defect number. Thus, for example, we list the various defects found in each of 20 fruitcakes, and find that the average or mean defect count is

$$\bar{c} = 5$$

Since the c chart is based on the Poisson distribution, the standard deviation is $s_c = \sqrt{c}$. Thus, to obtain limits at the usual 3 standard deviations

$$\text{U.C.L.}_c = \bar{c} + 3\sqrt{c} = 5 + 3\sqrt{5} = 11.71$$

$$\text{L.C.L.}_c = \bar{c} - 3\sqrt{c} = 5 - 3\sqrt{5} = -1.71, \text{ or } 0$$

No. OF DEFECTIVE
TOMATOES

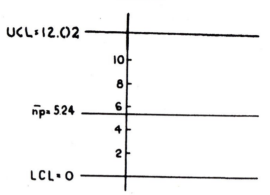

FIG. 19.6. CONTROL
CHART FOR ATTRIB-
UTES—np CHART

Using the data from Table
19.5, n = 200.

Thus, in our example, the c chart would have a vertical scale consisting of the number of defects per cake, with the central line ($\bar{c}$) set at 5, the upper control limit (U.C.L.$_c$) at 11.71, and the lower control limit (L.C.L.$_c$) at 0 (Fig. 19.7).

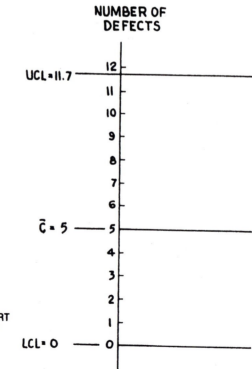

NUMBER OF
DEFECTS

FIG. 19.7. CONTROL CHART
FOR ATTRIBUTES—c CHART

For defects in fruitcakes.

REFERENCES

CLARK, C.T. and SCHKADE, L.L.   1969.   Statistical Methods for Business Decisions.   South-Western Publishing Co., Cincinnati.

DUNCAN, A.J.   1959.   Quality Control and Industrial Statistics.   Richard D. Irwin, Homewood, Ill.

ENRICK, N.L.   1965.   Quality Control and Reliability, 5th Edition.   Industrial Press, New York.

GRANT, E.L.   1952.   Statistical Quality Control, 2nd Edition.   McGraw-Hill Book Co., New York.

KORNETSKY, A. and KRAMER, A.   1957.   Quality control program for the processing of sweet corn. Food Technol. *11*, 188–192.

KRAMER, A. and TWIGG, B.A.   1970.   Quality Control for the Food Industry, 3rd Edition, Vol. 1.   AVI Publishing Co., Westport, Conn.

SHAININ, D.   1950.   The Hamilton Standard lot plot method of acceptance sampling by variables.   Ind. Qual. Control 7, 15–34.

SHEWHART, W.A.   1931.   Economic Control of Quality of Manufactured Products.   D. Van Nostrand Co., New York.

STOCKTON, J.R. and CLARK, C.T.   1971.   Introduction to Business and Economic Statistics, 4th Edition.   South-Western Publishing Co., Cincinnati.

20

Using Computers in Statistical Analysis

Whether you consider the computer to be the greatest invention since the wheel or the greatest curse visited upon mankind since the Black Plague, it is fairly safe to assume that the computer is here to stay. Properly used, computers can be a great aid in statistical analysis.

Every problem and example in this book can be performed with paper and pencil. In fact, every problem was initially calculated by hand. The larger problems were subsequently solved on the computer as a computational check.

The first step up from using paper and pencil is the use of a simple calculator. Many of the calculators manufactured during the 1950s and 1960s were specifically designed for statistical analysis—especially the calculation of sums of squares and cross products. With the introduction of the electronic calculator in the 1960s and 1970s, the ability to easily obtain sums of squares and cross products was lost. However, programmable calculators have reestablished the ability to quickly and easily calculate sums of squares and cross products. In addition, many of these programmable calculators have special subroutines which are preprogrammed for regression analysis.

If a statistical analysis is performed infrequently and fewer than 100 observations are involved, it is probably better to use a small handheld calculator or to perform the analysis with paper and pencil. On the other hand, if statistical analyses are to be performed frequently, or with many variables, or with thousands of observations, it will be worthwhile to become familiar with a computer package that can do the necessary routine calculations. Furthermore, many of these computer programs have the ability to draw histograms and scatter diagrams as a useful beginning point in any analysis.

Three common sets of statistical programs are:

BMDP—Biomedical Computer Programs
SAS—Statistical Analysis System
SPSS—Statistical Package for the Social Sciences

Each of these computer packages provides a detailed handbook describing each computer program, sample input, and sample output. Each has computer routines to plot data, perform the analysis of variance, multiple regression, and other statistical routines.

A good way to become familiar with a statistical program is to use it to solve a problem that you have previously solved by hand. In such a situation, since you already know the answer, you can devote your time to deciphering the computer printout. The first time that you solve a problem on the computer it will take considerably longer than doing the same problem by hand, but once you have been through the process, it will seem straightforward and become a great time saver. The real danger from using the computer is that it is such a great time saver that individuals become lazy and thoughtless. It is as easy to analyze 25 variables as 5. As a result, there is a tendency to "throw in every conceivable variable" and let the computer select those that are statistically significant. This tendency not only promotes muddled thinking, it also often yields misleading results.

At all times the computer must be made subordinate to the decision maker. There is no substitute for careful thought and the selection of appropriate statistical tools before turning to the computer as an aid to relieve the computational burden.

In order to illustrate the typical use of the BMDP computer package, Table 20.1 shows the data input for the problem shown in Table 13.9. The data of Table 20.1 are organized in the form of card input. They can be read as follows:

(1) Card 1
A run card which identifies the computer user.
(2) Card 2
An execution card which calls up the statistical program to be run.
(3) Card 3
A problem card which indicates that we have 42 observations, 3 original variables, that 9 transgeneration cards will be read and executed, and that 3 new variables will be created.
(4) Cards 4–6
These transgeneration cards create the 3 new variables that are needed. For example, card 4 creates the 4th variable (i.e., P^2) as the product (code 13) of the 1st variable (i.e., P) times the 1st variable.
(5) Cards 7–12
These cards scale all 6 variables as was done in Chapter 13. For example, card 7 defines a new 1st variable as the old 1st variable times (code 09) the constant 0.01.
(6) Card 13
This card defines the format for reading our data set.
(7) Cards 14–55
These cards report the original observations from Table 13.9.
(8) Card 56
This card states that the dependent variable is number 3.

TABLE 20.1.   COMPUTER DECK FOR THE DATA FROM TABLE 13.9

Card Number _____

Sheet Number _____

Line No.	
1	@RUN STAT,XXXZZZID
2	@XQT BMDUØMXLIBRARY.BMDO2R
3	PRØBLM TEST 42 3 9 3
4	TRNGEN 413 1 1
5	TRNGEN 513 2 2
6	TRNGEN 613 1 2
7	TRNGEN 109 1.01
8	TRNGEN 209 2.01
9	TRNGEN 309 3.01
10	TRNGEN 409 4.0001
11	TRNGEN 509 5.0001
12	TRNGEN 609 6.0001
13	(3F5.0)
14	60 90 73
15	60 90 75
16	60 100 82
17	60 100 80
18	60 110 85
49	100 120 100
50	100 130 109
51	100 130 110
52	100 140 117
53	100 140 120
54	100 150 125
55	100 150 127
56	SUBPRO 3
57	CONDEL221222
58	FINISH
59	@FIN

(9) Card 57

This card (the Control-Delete card) indicates that variables 1, 2, 4, 5, and 6 are to be used as explanatory variables.

(10) Cards 58–59

These cards terminate the run.

Table 20.2 shows the computer calculated regression results. Since the program used was a stepwise regression program (i.e., it calculated a different regression equation for each X brought into the equation), there were 5 steps calculated and reported. Only the final step is shown in Table 20.2, which is comparable to the results shown in Table 13.13. As can be seen by comparing the results in Table 20.2 with those of Table 13.13, some rounding errors occurred in our hand calculations in spite of our use of scaling and carrying 6 significant digits.

TABLE 20.2. COMPUTER PRINTED REGRESSION RESULTS FROM THE DATA IN TABLE 13.9

```
STEP NUMBER    5
VARIABLE ENTERED     1

MULTIPLE R            .9348
STD. ERROR OF EST.    .0511

ANALYSIS OF VARIANCE
                        DF    SUM OF SQUARES    MEAN SQUARE   F RATIO
           REGRESSION    5         .652            .130       49.886
           RESIDUAL     36         .094            .003

                    VARIABLES IN EQUATION
  VARIABLE         COEFFICIENT  STD. ERROR   F TO REMOVE

   (CONSTANT        .44245
     1            -1.49264      .73083       4.1724
     2             1.24524      .58088       4.5955
     4             -.35715      .41824       .7292
     5            -1.04365      .22767      21.0145
     6             2.04464      .24147      71.6961

F-LEVEL OR TOLERANCE INSUFFICIENT FOR FURTHER COMPUTATION
```

In addition to presenting the regression equation, the computer printout presents the summary ANOV table for this regression equation.

A second computer analysis is presented to illustrate the SAS computer package and a second mode of data input. For this example the data and program were entered on a typewriter-like terminal which transmitted to and received information from the computer over the telephone.

Although card input is not used in this example, instructions to the computer, the SAS program, and the data (Table 7.7) are submitted in card-images (card format). An overview of the complete process would be helpful before we consider the details of the SAS program and output.

The first step is to establish contact with the computer. This is accomplished by calling the computer and supplying it with the necessary identification and accounting information. When the computer is ready to accept information, a data file is created to provide a place to store the data of Table 7.7. The data are then typed into the file in card-image with the diet code typed in column 1 and the gain typed in columns 3–5. Similarly, a program file is created and the SAS program instructions are entered in card format. Since typographical errors and programming errors are not uncommon, editing procedures which permit the user to add, delete, or correct existing program or data lines are available.

Once the user is satisfied with the SAS program and data files, he is ready to submit his program to the computer for execution. This is accomplished by typing in instructions to allocate the 2 files so that the computer will know which files are to be used and then to type an execution command.

Now let us turn our attention to the SAS program used to analyze the data of Table 7.7. A data list is given in Table 20.3 as it appeared at execution of the program. The actual lines in the program file are identified by the 5 digit number preceding each line, numbered from 10 to 70 in tens, making 7 lines in all. This number is provided by the computer. A detailed description of the syntax can be found in the SAS Handbook. We will direct our attention to the function of each line (Table 20.3).

TABLE 20.3.   PROGRAM FILE FOR ANALYSIS OF THE DATA FROM TABLE 7.7

```
LIST STEERS.CNTL
 STEERS.CNTL
00010 DATA TABLE77;
00020    INFILE A;
00030    INPUT DIET GAIN;
00040 PROC ANOVA;
00050    CLASSES DIET;
00060    MODEL GAIN = DIET;
00070    MEANS DIET / DUNCAN;
READY
```

(1) Line #10
Assigns the name TABLE77 to the SAS data set to be processed.

(2) Line #20
Identifies the data set as being allocated to file A permitting the computer to locate the data file.

(3) Line #30
Identifies that diet is in the first field and that gain is in the second field of the data file.

(4) Line #40
A statement indicating that the desired data processing procedure is the ANOVA.

(5) Line #50
Identifies diet as the classification variable.

(6) Line #60
This is a model statement similar to those presented in the ANOVA chapters. The μ and ϵ terms are assumed by the SAS ANOVA and are not included in the model. The statement identifies gain as the Y variable (dependent) and diet as the treatment variable (independent).

(7) Line #70
This line requests that means be computed and a Duncan's multiple range test be conducted.

Table 20.4 is the listing of the data file. Each data line is again preceded by a computer-provided identification number. The actual data input is diet in column 1, a blank in column 2, and gain in columns 3–5.

TABLE 20.4. DATA FILE FOR THE DATA FROM TABLE 7.7

```
LIST STEERS.DATA
  STEERS.DATA
00010 1   91
00020 1   89
00030 1  102
00040 1   82
00050 1   93
00060 1   93
00070 2  110
00080 2   84
00090 2   84
00100 2   85
00110 2   89
00120 2   87
00130 3   85
00140 3   92
00150 3   89
00160 3   85
00170 3   72
00180 3   80
00190 4   80
00200 4   72
00210 4   89
00220 4   79
00230 4   72
00240 4   76
READY
```

The output from the SAS program is given in Table 20.5. Computed quantities are well labeled and should be familiar to you. Due to rounding errors the computer and hand calculated values will not necessarily agree exactly. You may wish to compare this value with the corresponding values calculated in Chapter 7.

Tables 20.6 through 20.11 are additional examples of outputs from SAS programs on data sets used as examples in Chapters 7 through 11.

TABLE 20.5. SAS ANOVA AND DUNCAN'S COMPUTER OUTPUT FOR THE DATA FROM TABLE 7.7

ANALYSIS OF VARIANCE PROCEDURE

CLASS LEVEL INFORMATION

CLASS	LEVELS	VALUES
DIET	4	1 2 3 4

NUMBER OF OBSERVATIONS IN DATA SET = 24

ANALYSIS OF VARIANCE PROCEDURE

DEPENDENT VARIABLE: GAIN

SOURCE	DF	SUM OF SQUARES	MEAN SQUARE
MODEL	3	692.33333333	230.77777778
ERROR	20	1171.00000000	58.55000000
CORRECTED TOTAL	23	1863.33333333	

MODEL F =	3.94		PR > F = 0.0232

R-SQUARE	C.V.	STD DEV	GAIN MEAN
0.371556	8.9147	7.65179717	85.83333333

SOURCE	DF	ANOVA SS	F VALUE	PR > F
DIET	3	692.33333333	3.94	0.0232

ANALYSIS OF VARIANCE PROCEDURE

DUNCAN'S MULTIPLE RANGE TEST FOR VARIABLE GAIN

MEANS WITH THE SAME LETTER ARE NOT SIGNIFICANTLY DIFFERENT.

ALPHA LEVEL=.05   DF=20   MS=58.55

GROUPING	MEAN	N	DIET
A	91.666667	6	1
A			
A	89.833333	6	2
A			
B A	83.833333	6	3
B			
B	78.000000	6	4

TABLE 20.6.   SAS NESTED ANOVA FOR THE DATA FROM TABLE 7.10

COEFFICIENTS OF EXPECTED MEAN SQUARES

SOURCE	SIRE	ERROR
SIRE	3.65385	1.00000
ERROR	0.0	1.00000

ANALYSIS OF VARIABLE HANDS

VARIANCE SOURCE	D.F.	SUM OF SQUARES	MEAN SQUARES	VARIANCE COMPONENT	PERCENT
TOTAL	25	37.65385	1.50615	1.57150	100.00
SIRE	6	17.88301	2.98050	0.53093	33.78
ERROR	19	19.77083	1.04057	1.04057	66.22
MEAN			11.884615		
STANDARD DEVIATION			1.020083		
COEFFICIENT OF VARIATION			0.085832		

TABLE 20.7.   SAS ANOVA AND DUNCAN'S FOR THE DATA FROM TABLE 8.1

ANALYSIS OF VARIANCE PROCEDURE

CLASS LEVEL INFORMATION

CLASS	LEVELS	VALUES
VARIETY	6	A B C D E F
BLOCK	4	1 2 3 4

NUMBER OF OBSERVATIONS IN DATA SET = 24

ANALYSIS OF VARIANCE PROCEDURE

DEPENDENT VARIABLE: YIELD

SOURCE	DF	SUM OF SQUARES	MEAN SQUARE
MODEL	8	1954.66666667	244.33333333
ERROR	15	547.33333333	36.48888889
CORRECTED TOTAL	23	2502.00000000	

MODEL F =	6.70			PR > F = 0.0008

R-SQUARE	C.V.	STD DEV	YIELD MEAN
0.781242	22.3726	6.04060335	27.00000000

SOURCE	DF	ANOVA SS	F VALUE	PR > F
VARIETY	5	711.00000000	3.90	0.0183
BLOCK	3	1243.66666667	11.36	0.0004

ANALYSIS OF VARIANCE PROCEDURE

DUNCAN'S MULTIPLE RANGE TEST FOR VARIABLE YIELD

MEANS WITH THE SAME LETTER ARE NOT SIGNIFICANTLY DIFFERENT.

ALPHA LEVEL=.05   DF=15   MS=36.4889

GROUPING			MEAN	N	VARIETY
	A		33.750000	4	B
	A				
B	A		31.750000	4	A
B	A				
B	A		30.250000	4	C
B	A				
B	A	C	25.000000	4	D
B		C			
B		C	23.250000	4	E
B		C			
		C	18.000000	4	F

TABLE 20.8.   SAS ANOVA FOR THE DATA FROM TABLE 8.8

ANALYSIS OF VARIANCE PROCEDURE

CLASS LEVEL INFORMATION

CLASS	LEVELS	VALUES
TRT	4	A B C D
PERIOD	4	I II III IV
CHAMBER	4	1 2 3 4

NUMBER OF OBSERVATIONS IN DATA SET = 16

ANALYSIS OF VARIANCE PROCEDURE

DEPENDENT VARIABLE: DM

SOURCE	DF	SUM OF SQUARES	MEAN SQUARE
MODEL	9	18446.50000000	2049.61111111
ERROR	6	2799.50000000	466.58333333
CORRECTED TOTAL	15	21246.00000000	

MODEL F =   4.39   PR > F = 0.0428

R-SQUARE	C.V.	STD DEV	DM MEAN
0.868234	5.9588	21.60054012	362.50000000

SOURCE	DF	ANOVA SS	F VALUE	PR > F
TRT	3	459.50000000	0.33	0.8056
PERIOD	3	6689.50000000	4.78	0.0496
CHAMBER	3	11298.50000000	8.07	0.0158

TABLE 20.9.   SAS ANOVA FOR THE DATA FROM TABLE 9.4

```
ANALYSIS OF VARIANCE PROCEDURE

CLASS LEVEL INFORMATION

CLASS      LEVELS     VALUES

PROTEIN      2       HIGH LOW

FAT          2       HIGH LOW


NUMBER OF OBSERVATIONS IN DATA SET = 28

ANALYSIS OF VARIANCE PROCEDURE

DEPENDENT VARIABLE: INTAKE
```

SOURCE	DF	SUM OF SQUARES	MEAN SQUARE
MODEL	3	1009.25000000	336.41666667
ERROR	24	1885.42857143	78.55952381
CORRECTED TOTAL	27	2894.67857143	

MODEL F =	4.28		PR F = 0.0148

R-SQUARE	C.V.	STD DEV	INTAKE MEAN
0.348657	20.3255	8.86338106	43.60714286

SOURCE	DF	ANOVA SS	F VALUE	PR F
PROTEIN	1	558.03571429	7.10	0.0135
FAT	1	72.32142857	0.92	0.3469
PROTEIN*FAT	1	378.89285714	4.82	0.0380

```
ANALYSIS OF VARIANCE PROCEDURE

MEANS
```

PROTEIN	N	INTAKE
HIGH	14	39.1428571
LOW	14	48.0714286

FAT	N	INTAKE
HIGH	14	42.0000000
LOW	14	45.2142857

PROTEIN	FAT	N	INTAKE
HIGH	HIGH	7	33.8571429
HIGH	LOW	7	44.4285714
LOW	HIGH	7	50.1428571
LOW	LOW	7	46.0000000

TABLE 20.10.   SAS ANOCOVA AND ADJUSTED MEANS FOR THE DATA FROM TABLE 10.1

GENERAL LINEAR MODELS PROCEDURE

CLASS LEVEL INFORMATION

CLASS	LEVELS	VALUES
BLOCK	5	1 2 3 4 5
TRT	2	A B

NUMBER OF OBSERVATIONS IN DATA SET = 10

GENERAL LINEAR MODELS PROCEDURE

DEPENDENT VARIABLE: FINIAL

SOURCE	DF	SUM OF SQUARES	MEAN SQUARE
MODEL	6	0.50877159	0.08479526
ERROR	3	0.06387841	0.02129280
CORRECTED TOTAL	9	0.57265000	

MODEL F =	3.98		PR > F = 0.1420

R-SQUARE	C.V.	STD DEV	FINIAL MEAN
0.888451	10.5358	0.14592054	1.38500000

SOURCE	DF	TYPE IV SS	F VALUE	PR > F
BLOCK	4	0.05990532	0.70	0.6398
TRT	1	0.15838611	7.44	0.0721
INITIAL	1	0.19532159	9.17	0.0564

GENERAL LINEAR MODELS PROCEDURE

LEAST SQUARES MEANS

TRT	FINIAL LSMEAN	STD ERR LSMEAN	PROB > |T| H0:LSMEAN=0
A	1.59073379	0.08842779	0.0004
B	1.17926621	0.08842779	0.0009

TABLE 20.11.   SAS CONTINGENCY TABLE ANALYSIS FOR THE DATA FROM TABLE 11.3

```
TABLE OF REGIME BY PROGRESS
REGIME          PROGRESS

FREQUENCY!
EXPECTED  !
DEVIATION!
CELL CHI2!
  ROW PCT !+          !++         !-0         !  TOTAL
---------+-----------+-----------+-----------+
I        !        9 !        6 !       15 !    30
         !     13.0 !      6.7 !     10.3 !
         !     -4.0 !     -0.7 !      4.7 !
         !      1.2 !      0.1 !      2.1 !
         !    30.00 !    20.00 !    50.00 !
---------+-----------+-----------+-----------+
II       !       15 !       11 !        4 !    30
         !     13.0 !      6.7 !     10.3 !
         !      2.0 !      4.3 !     -6.3 !
         !      0.3 !      2.8 !      3.9 !
         !    50.00 !    36.67 !    13.33 !
---------+-----------+-----------+-----------+
III      !       15 !        3 !       12 !    30
         !     13.0 !      6.7 !     10.3 !
         !      2.0 !     -3.7 !      1.7 !
         !      0.3 !      2.0 !      0.3 !
         !    50.00 !    10.00 !    40.00 !
---------+-----------+-----------+-----------+
TOTAL            39         20         31        90
STATISTICS FOR 2-WAY TABLES
CHI-SQUARE                           13.004   DF=    4   PROB=0.0113
PHI                                   0.380
CONTINGENCY COEFFICIENT               0.355
CRAMER'S V                            0.269
LIKELIHOOD RATIO CHISQUARE           14.105   DF=    4   PROB=0.0070
```

REFERENCES

Many of the small programmable calculators offer preprogrammed modules with detailed instructions to perform many of the standard statistical calculations. For example:

Applied Statistics
TI Programmable 58/59
Texas Instruments, Inc.
Dallas, Texas

Large scale computers usually have a library of computer packages which will perform standard statistical analyses. Each computer manufacturer offers such programs. In addition, many computer centers will have one or more of the following available:

BMDP-77
Biomedical Computer Programs, P-Series
University of California Press
University of California, Los Angeles

SAS
Statistical Analysis System
A User's Guide to SAS-79
SAS Institute
P.O. Box 10066
Raleigh, North Carolina

SPSS
Statistical Package for the Social Sciences
McGraw-Hill Book Co.
New York, N.Y.

Appendix

TABLE A.1. SOURCES OF SECONDARY DATA

In general the food scientist works with primary data. However, from time to time secondary data can play an important role in placing existing research or experience in its proper historical or geographic context. In addition, secondary data may often be analyzed directly—particularly within an economic framework.

Although nearly all state governments publish agricultural statistics, the main source of secondary data is the federal government. The following documents are usually available in public or university libraries. They can also be obtained directly from the appropriate agencies free or at a nominal cost.

For broad aggregate statistics the following are useful:

(1) Agricultural Statistics
     (Updated annually by the U.S. Dept. of Agriculture)
     For sale by the:
        Superintendent of Documents
        U.S. Government Printing Office
        Washington, D.C. 20402

(2) Statistical Abstract of the United States
     (Updated annually by the U.S. Dept. of Commerce)
     For sale by the:
        Superintendent of Documents
        U.S. Government Printing Office
        Washington, D.C. 20402

For current statistics the following are useful:

(1) Survey of Current Business
     (Published monthly by the Bureau of Economic Analysis of the U.S. Dept. of Commerce)

For sale by the:
Superintendent of Documents
U.S. Government Printing Office
Washington, D.C. 20402

(2)   Commodity Situation Reports are issued for Cotton and Wool; Dairy; Fats and Oils; Feed; Fruits; Livestock and Meat; National Food; Poultry and Egg; Rice; Sugar and Sweetener; Tobacco; Vegetable; and Wheat. These commodity reports analyze supply and demand, price, and the outlook for major farm commodities. Tables and charts present current data on production, market movement, stocks, consumption, prices, and foreign trade.

For more information write:
U.S. Department of Agriculture
ESS Information Staff
Publications Unit, Room 0054-South
Washington, D.C. 20250

(3)   Additional data are available in the form of statistical series on Crops; Livestock and Poultry; Dairy; Price; and Other (Cold Storage, Farm Labor, Honey, etc.). These are released at various times throughout the year. Some reports are issued monthly. For more information write:

Crop Reporting Board
Economics and Statistics Service
U.S. Department of Agriculture
Washington, D.C. 20250

The report *Scope and Methods of the Statistical Reporting Service*, Misc. Publ. *1308*, U.S. Dept. Agric., provides a detailed presentation of the data collected and the statistical procedures used by the USDA.

(4)   Statistical compilations of per capita production, consumption, carry-overs, etc., are frequently obtainable from trade periodicals or trade association reports. Some examples are:

(a)   The Almanac of the Canning, Freezing, Preserving Industries. Updated annually by Edward E. Judge & Sons, Westminster, Md.

(b)   Cold Storage Reports. Weekly-Monthly, by American Frozen Food Institute, McLean, Va.

(c)   Modern Packaging. Encyclopedia issue, 13th annual issue. Published by McGraw-Hill, New York.

TABLE B.1.   AREAS UNDER THE STANDARD NORMAL CURVE

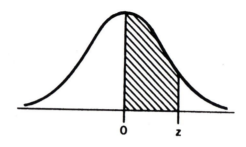

The following table provides the area between the mean and normal deviate value Z.

Normal Deviate Z	0.00	0.01	0.02	0.03	0.04	0.05	0.06	0.07	0.08	0.09
0.0	0.0000	0.0040	0.0080	0.0120	0.0160	0.0199	0.0239	0.0279	0.0319	0.0359
0.1	0.0398	0.0438	0.0478	0.0517	0.0557	0.0596	0.0636	0.0675	0.0714	0.0753
0.2	0.0793	0.0832	0.0871	0.0910	0.0948	0.0987	0.1026	0.1064	0.1103	0.1141
0.3	0.1179	0.1217	0.1255	0.1293	0.1331	0.1368	0.1406	0.1443	0.1480	0.1517
0.4	0.1554	0.1591	0.1628	0.1664	0.1700	0.1736	0.1772	0.1808	0.1844	0.1879
0.5	0.1915	0.1950	0.1985	0.2019	0.2054	0.2088	0.2123	0.2157	0.2190	0.2224
0.6	0.2257	0.2291	0.2324	0.2357	0.2389	0.2422	0.2454	0.2486	0.2518	0.2549
0.7	0.2580	0.2612	0.2642	0.2673	0.2704	0.2734	0.2764	0.2794	0.2823	0.2852
0.8	0.2881	0.2910	0.2939	0.2967	0.2995	0.3023	0.3051	0.3078	0.3106	0.3133
0.9	0.3159	0.3186	0.3212	0.3238	0.3264	0.3289	0.3315	0.3340	0.3365	0.3389
1.0	0.3413	0.3438	0.3461	0.3485	0.3508	0.3531	0.3554	0.3577	0.3599	0.3621
1.1	0.3643	0.3665	0.3686	0.3708	0.3729	0.3749	0.3770	0.3790	0.3810	0.3830
1.2	0.3849	0.3869	0.3888	0.3907	0.3925	0.3944	0.3962	0.3980	0.3997	0.4015
1.3	0.4032	0.4049	0.4066	0.4082	0.4099	0.4115	0.4131	0.4147	0.4162	0.4177
1.4	0.4192	0.4207	0.4222	0.4236	0.4251	0.4265	0.4279	0.4292	0.4306	0.4319
1.5	0.4332	0.4345	0.4357	0.4370	0.4382	0.4394	0.4406	0.4418	0.4429	0.4441
1.6	0.4452	0.4463	0.4474	0.4484	0.4495	0.4505	0.4515	0.4525	0.4535	0.4545
1.7	0.4554	0.4564	0.4573	0.4582	0.4591	0.4599	0.4608	0.4616	0.4625	0.4633
1.8	0.4641	0.4649	0.4656	0.4664	0.4671	0.4678	0.4686	0.4693	0.4699	0.4706
1.9	0.4713	0.4719	0.4726	0.4732	0.4738	0.4744	0.4750	0.4756	0.4761	0.4767
2.0	0.4772	0.4778	0.4783	0.4788	0.4793	0.4798	0.4803	0.4808	0.4812	0.4817
2.1	0.4821	0.4826	0.4830	0.4834	0.4838	0.4842	0.4846	0.4850	0.4854	0.4857
2.2	0.4861	0.4864	0.4868	0.4871	0.4875	0.4878	0.4881	0.4884	0.4887	0.4890
2.3	0.4893	0.4896	0.4898	0.4901	0.4904	0.4906	0.4909	0.4911	0.4913	0.4916
2.4	0.4918	0.4920	0.4922	0.4925	0.4927	0.4929	0.4931	0.4932	0.4934	0.4936
2.5	0.4938	0.4940	0.4941	0.4943	0.4945	0.4946	0.4948	0.4949	0.4951	0.4952
2.6	0.4953	0.4955	0.4956	0.4957	0.4959	0.4960	0.4961	0.4962	0.4963	0.4964
2.7	0.4965	0.4966	0.4967	0.4968	0.4969	0.4970	0.4971	0.4972	0.4973	0.4974
2.8	0.4974	0.4975	0.4976	0.4977	0.4977	0.4978	0.4979	0.4979	0.4980	0.4981
2.9	0.4981	0.4982	0.4982	0.4983	0.4984	0.4984	0.4985	0.4985	0.4986	0.4986
3.0	0.49865	0.4987	0.4987	0.4988	0.4988	0.4989	0.4989	0.4989	0.4990	0.4990
4.0	0.49997									

Source: Hamburg (1970). (Reproduced by permission)

TABLE C.1.   THE DISTRIBUTION OF t

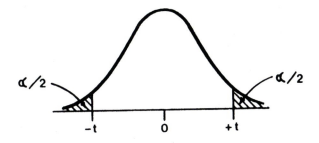

Degrees of Freedom (df)	Probability of a Larger Value of t, Sign Ignored					Degrees of Freedom (df)
	0.5	0.1	0.05	0.02	0.01	
1	1.000	6.314	12.706	31.821	63.657	1
2	0.816	2.920	4.303	6.965	9.925	2
3	0.765	2.353	3.182	4.541	5.841	3
4	0.741	2.132	2.776	3.747	4.604	4
5	0.727	2.015	2.571	3.365	4.032	5
6	0.718	1.943	2.447	3.143	3.707	6
7	1.711	1.895	2.365	2.998	3.499	7
8	0.706	1.860	2.306	2.896	3.355	8
9	0.703	1.833	2.262	2.821	3.250	9
10	0.700	1.812	2.228	2.764	3.169	10
11	0.697	1.796	2.201	2.718	3.106	11
12	1.695	1.782	2.179	2.681	3.055	12
13	0.694	1.771	2.160	2.650	3.012	13
14	0.692	1.761	2.145	2.624	2.977	14
15	0.691	1.753	2.131	2.602	2.947	15
16	0.690	1.746	2.120	2.583	2.921	16
17	0.689	1.740	2.110	2.567	2.898	17
18	0.688	1.734	2.101	2.552	2.878	18
19	0.688	1.729	2.093	2.539	2.861	19
20	0.687	1.725	2.086	2.528	2.845	20
21	0.686	1.721	2.080	2.518	2.831	21
22	0.686	1.717	2.074	2.508	2.819	22
23	0.685	1.714	2.069	2.500	2.807	23
24	0.685	1.711	2.064	2.492	2.797	24
25	0.684	1.708	2.060	2.485	2.787	25
26	0.684	1.706	2.056	2.479	2.779	26
27	0.684	1.703	2.052	2.473	2.771	27
28	0.683	1.701	2.048	2.467	2.763	28
29	0.683	1.699	2.045	2.462	2.756	29
30	0.683	1.697	2.042	2.457	2.750	30
35	—	—	2.030	—	2.724	35
40	—	—	2.021	—	2.704	40
45	—	—	2.014	—	2.690	45
50	—	—	2.008	—	2.678	50
60	—	—	2.000	—	2.660	60
80	—	—	1.990	—	2.638	80
100	—	—	1.984	—	2.626	100
200	—	—	1.972	—	2.601	200
500	—	—	1.965	—	2.568	500
1000	—	—	1.962	—	2.581	1000
∞	0.67449	1.64485	1.95996	2.32634	2.57582	∞

Source: Fisher (1925). (Reproduced by permission of author and publishers)

TABLE D.1. VALUES OF F FOR THE 5% LEVEL OF SIGNIFICANCE

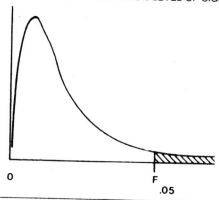

0   F .05

Numerator—Degrees of Freedom

m\n	1	2	3	4	5	6	7	8	9
1	161.4	199.5	215.7	224.6	230.2	234.0	238.8	238.9	240.5
2	18.51	19.00	19.16	19.25	19.30	19.33	19.35	19.37	19.38
3	10.13	9.55	9.28	9.12	9.01	8.94	8.89	8.85	8.81
4	7.71	6.94	6.59	6.39	6.26	6.16	6.09	6.04	6.00
5	6.61	5.79	5.41	5.19	5.05	4.95	4.88	4.82	4.77
6	5.99	5.14	4.76	4.53	4.39	4.28	4.21	4.15	4.10
7	5.59	4.74	4.35	4.12	3.97	3.87	3.79	3.73	3.68
8	5.32	4.46	4.07	3.84	3.69	3.58	3.50	3.44	3.39
9	5.12	4.26	3.86	3.63	3.48	3.37	3.29	3.23	3.18
10	4.96	4.10	3.71	3.48	3.33	3.22	3.14	3.07	3.02
11	4.84	3.98	3.59	3.36	3.20	3.09	3.01	2.95	2.90
12	4.75	3.89	3.49	3.26	3.11	3.00	2.91	2.85	2.80
13	4.67	3.81	3.41	3.18	3.03	2.92	2.83	2.77	2.71
14	4.60	3.74	3.34	3.11	2.96	2.85	2.76	2.70	2.65
15	4.54	3.68	3.29	3.06	2.90	2.79	2.71	2.64	2.59
16	4.49	3.63	3.24	3.01	2.85	2.74	2.66	2.59	2.54
17	4.45	3.59	3.20	2.96	2.81	2.70	2.61	2.55	2.49
18	4.41	3.55	3.16	2.93	2.77	2.66	2.58	2.51	2.46
19	4.38	3.52	3.13	2.90	2.74	2.63	2.54	2.48	2.42
20	4.35	3.49	3.10	2.87	2.71	2.60	2.51	2.45	2.39
21	4.32	3.47	3.07	2.84	2.68	2.57	2.49	2.42	2.37
22	4.30	3.44	3.05	2.82	2.66	2.55	2.46	2.40	2.34
23	4.28	3.42	3.03	2.80	2.64	2.53	2.44	2.37	2.32
24	4.26	3.40	3.01	2.78	2.62	2.51	2.42	2.36	2.30
25	4.24	3.39	2.99	2.76	2.60	2.49	2.40	2.34	2.28
26	4.23	3.37	2.98	2.74	2.59	2.47	2.39	2.32	2.27
27	4.21	3.35	2.96	2.73	2.57	2.46	2.37	2.31	2.25
28	4.20	3.34	2.95	2.71	2.56	2.45	2.36	2.29	2.24
29	4.18	3.33	2.93	2.70	2.55	2.43	2.35	2.28	2.22
30	4.17	3.32	2.92	2.69	2.53	2.42	2.33	2.27	2.21
40	4.08	3.23	2.84	2.61	2.45	2.34	2.25	2.18	2.12
60	4.00	3.15	2.76	2.53	2.37	2.25	2.17	2.10	2.04
120	3.92	3.07	2.68	2.45	2.29	2.17	2.09	2.02	1.96
∞	3.84	3.00	2.60	2.37	2.21	2.10	2.01	1.94	1.88

Denominator—Degrees of Freedom

Source: Hickman and Hilton (1971).

$$F = \frac{s_1^2}{s_2^2} = \frac{S_1}{m} \Big/ \frac{S_2}{n}, \text{ where } s_1^2 = S_1/m \text{ and } s_2^2 = S_2/n \text{ are independent mean squares}$$

Numerator—Degrees of Freedom									
10	12	15	20	24	30	40	60	120	∞
241.9	243.9	245.9	248.0	249.1	250.1	251.1	252.2	253.3	254.3
19.40	19.41	19.43	19.45	19.45	19.46	19.47	19.48	19.49	19.50
8.79	8.74	8.70	8.66	8.64	8.62	8.59	8.57	8.55	8.53
5.96	5.91	5.86	5.80	5.77	5.75	5.72	5.69	5.66	5.63
4.74	4.68	4.62	4.56	4.53	4.50	4.46	4.43	4.40	4.36
4.06	4.00	3.94	3.87	3.84	3.81	3.77	3.74	3.70	3.67
3.64	3.57	3.51	3.44	3.41	3.38	3.34	3.30	3.27	3.23
3.35	3.28	3.22	3.15	3.12	3.08	3.04	3.01	2.97	2.93
3.14	3.07	3.01	2.94	2.90	2.86	2.83	2.79	2.75	2.71
2.98	2.91	2.85	2.77	2.74	2.70	2.66	2.62	2.58	2.54
2.85	2.79	2.72	2.65	2.61	2.57	2.53	2.49	2.45	2.40
2.75	2.69	2.62	2.54	2.51	2.47	2.43	2.38	2.34	2.30
2.67	2.60	2.53	2.46	2.42	2.38	2.34	2.30	2.25	2.21
2.60	2.53	2.46	2.39	2.35	2.31	2.27	2.22	2.18	2.13
2.54	2.48	2.40	2.33	2.29	2.25	2.20	2.16	2.11	2.07
2.49	2.42	2.35	2.28	2.24	2.19	2.15	2.11	2.06	2.01
2.45	2.38	2.31	2.23	2.19	2.15	2.10	2.06	2.01	1.96
2.41	2.34	2.27	2.19	2.15	2.11	2.06	2.02	1.97	1.92
2.38	2.31	2.23	2.16	2.11	2.07	2.03	1.98	1.93	1.88
2.35	2.28	2.20	2.12	2.08	2.04	1.99	1.95	1.90	1.84
2.32	2.25	2.18	2.10	2.05	2.01	1.96	1.92	1.87	1.81
2.30	2.23	2.15	2.07	2.03	1.98	1.94	1.89	1.84	1.78
2.27	2.20	2.13	2.05	2.01	1.96	1.91	1.86	1.81	1.76
2.25	2.18	2.11	2.03	1.98	1.94	1.89	1.84	1.79	1.73
2.24	2.16	2.09	2.01	1.96	1.92	1.87	1.82	1.77	1.71
2.22	2.15	2.07	1.99	1.95	1.90	1.85	1.80	1.75	1.69
2.20	2.13	2.06	1.97	1.93	1.88	1.84	1.79	1.73	1.67
2.19	2.12	2.04	1.96	1.91	1.87	1.82	1.77	1.71	1.65
2.18	2.10	2.03	1.94	1.90	1.85	1.81	1.75	1.70	1.64
2.16	2.09	2.01	1.93	1.89	1.84	1.79	1.74	1.68	1.62
2.08	2.00	1.92	1.84	1.79	1.74	1.69	1.64	1.58	1.51
1.99	1.92	1.84	1.76	1.70	1.65	1.59	1.53	1.47	1.39
1.91	1.83	1.75	1.66	1.61	1.55	1.50	1.43	1.35	1.26
1.83	1.75	1.67	1.57	1.52	1.46	1.39	1.32	1.22	1.00

estimating a common variance σ^2 and based on m and n degrees of freedom, respectively.

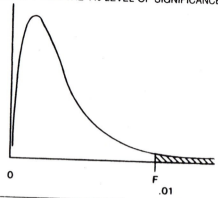

		Numerator—Degrees of Freedom							
n \ m	1	2	3	4	5	6	7	8	9
1	4052	4999.5	5403	5625	5764	5859	5928	5982	6022
2	98.50	99.00	99.17	99.25	99.30	99.33	99.36	99.37	99.39
3	34.12	30.82	29.46	28.71	28.24	27.91	27.67	27.49	27.35
4	21.20	18.00	16.69	15.98	15.52	15.21	14.98	14.80	14.66
5	16.26	13.27	12.06	11.39	10.97	10.67	10.46	10.29	10.16
6	13.75	10.92	9.78	9.15	8.75	8.47	8.26	8.10	7.98
7	12.25	9.55	8.45	7.85	7.46	7.19	6.99	6.84	6.72
8	11.26	8.65	7.59	7.01	6.63	6.37	6.18	6.03	5.91
9	10.56	8.02	6.99	6.42	6.06	5.80	5.61	5.47	5.35
10	10.04	7.56	6.55	5.99	5.64	5.39	5.20	5.06	4.94
11	9.65	7.21	6.22	5.67	5.32	5.07	4.89	4.74	4.63
12	9.33	6.93	5.95	5.41	5.06	4.82	4.64	4.50	4.39
13	9.07	6.70	5.74	5.21	4.86	4.62	4.44	4.30	4.19
14	8.86	6.51	5.56	5.04	4.69	4.46	4.28	4.14	4.03
15	8.68	6.36	5.42	4.89	4.56	4.32	4.14	4.00	3.89
16	8.53	6.23	5.29	4.77	4.44	4.20	4.03	3.89	3.78
17	8.40	6.11	5.18	4.67	4.34	4.10	3.93	3.79	3.68
18	8.29	6.01	5.09	4.58	4.25	4.01	3.84	3.71	3.60
19	8.18	5.93	5.01	4.50	4.17	3.94	3.77	3.63	3.52
20	8.10	5.85	4.94	4.43	4.10	3.87	3.70	3.56	3.46
21	8.02	5.78	4.87	4.37	4.04	3.81	3.64	3.51	3.40
22	7.95	5.72	4.82	4.31	3.99	3.76	3.59	3.45	3.35
23	7.88	5.66	4.76	4.26	3.94	3.71	3.54	3.41	3.30
24	7.82	5.61	4.72	4.22	3.90	3.67	3.50	3.36	3.26
25	7.77	5.57	4.68	4.18	3.85	3.63	3.46	3.32	3.22
26	7.72	5.53	4.64	4.14	3.82	3.59	3.42	3.29	3.18
27	7.68	5.49	4.60	4.11	3.78	3.56	3.39	3.26	3.15
28	7.64	5.45	4.57	4.07	3.75	3.53	3.36	3.23	3.12
29	7.60	5.42	4.54	4.04	3.73	3.50	3.33	3.20	3.09
30	7.56	5.39	4.51	4.02	3.70	3.47	3.30	3.17	3.07
40	7.31	5.18	4.31	3.83	3.51	3.29	3.12	2.99	2.89
60	7.08	4.98	4.13	3.65	3.34	3.12	2.95	2.82	2.72
120	6.85	4.79	3.95	3.48	3.17	2.96	2.79	2.66	2.56
∞	6.63	4.61	3.78	3.32	3.02	2.80	2.64	2.51	2.41

Source: Hickman and Hilton (1971).

$$F = \frac{s_1^2}{s_2^2} = \frac{S_1}{m} \Big/ \frac{S_2}{n}, \text{ where } s_1^2 = S_1/m \text{ and } s_2^2 = S_2/n \text{ are independent mean squares estimating}$$

			Numerator—Degrees of Freedom						
10	12	15	20	24	30	40	60	120	∞
6056	6106	6157	6209	6235	6261	6287	6313	6339	6366
99.40	99.42	99.43	99.45	99.46	99.47	99.47	99.48	99.49	99.50
27.23	27.05	26.87	26.69	26.60	26.50	26.41	26.32	26.22	26.13
14.55	14.37	14.20	14.02	13.93	13.84	13.75	13.65	13.56	13.46
10.05	9.89	9.72	9.55	9.47	9.38	9.29	9.20	9.11	9.02
7.87	7.72	7.56	7.40	7.31	7.23	7.14	7.06	6.97	6.88
6.62	6.47	6.31	6.16	6.07	5.99	5.91	5.82	5.74	5.65
5.81	5.67	5.52	5.36	5.28	5.20	5.12	5.03	4.95	4.86
5.26	5.11	4.96	4.81	4.73	4.65	4.57	4.48	4.40	4.31
4.85	4.71	4.56	4.41	4.33	4.25	4.17	4.08	4.00	3.91
4.54	4.40	4.25	4.10	4.02	3.94	3.86	3.78	3.69	3.60
4.30	4.16	4.01	3.86	3.78	3.70	3.62	3.54	3.45	3.36
4.10	3.96	3.82	3.66	3.59	3.51	3.43	3.34	3.25	3.17
3.94	3.80	3.66	3.51	3.43	3.35	3.27	3.18	3.09	3.00
3.80	3.67	3.52	3.37	3.29	3.21	3.13	3.05	2.96	2.87
3.69	3.55	3.41	3.26	3.18	3.10	3.02	2.93	2.84	2.75
3.59	3.46	3.31	3.16	3.08	3.00	2.92	2.83	2.75	2.65
3.51	3.37	3.23	3.08	3.00	2.92	2.84	2.75	2.65	2.57
3.43	3.30	3.15	3.00	2.92	2.84	2.76	2.67	2.58	2.49
3.37	3.23	3.09	2.94	2.86	2.78	2.69	2.61	2.52	2.42
3.31	3.17	3.03	2.88	2.80	2.72	2.64	2.55	2.46	2.36
3.26	3.12	2.98	2.83	2.75	2.67	2.58	2.50	2.40	2.31
3.21	3.07	2.93	2.78	2.70	2.62	2.54	2.45	2.35	2.26
3.17	3.03	2.89	2.74	2.66	2.58	2.49	2.40	2.31	2.21
3.13	2.99	2.85	2.70	2.62	2.54	2.45	2.36	2.27	2.17
3.09	2.96	2.81	2.66	2.58	2.50	2.42	2.33	2.23	2.13
3.06	2.93	2.78	2.63	2.55	2.47	2.38	2.29	2.20	2.10
3.03	2.90	2.75	2.60	2.52	2.44	2.35	2.26	2.17	2.06
3.00	2.87	2.73	2.57	2.49	2.41	2.33	2.23	2.14	2.03
2.98	2.84	2.70	2.55	2.47	2.39	2.30	2.21	2.11	2.01
2.80	2.66	2.52	2.37	2.29	2.20	2.11	2.02	1.92	1.80
2.63	2.50	2.35	2.20	2.12	2.03	1.94	1.84	1.73	1.60
2.47	2.34	2.19	2.03	1.90	1.86	1.76	1.76	1.53	1.38
2.32	2.18	2.04	1.88	1.79	1.70	1.59	1.47	1.32	1.00

a common variance σ^2 and based on m and n degrees of freedom, respectively.

TABLE F.1. COEFFICIENTS AND DIVISORS FOR SETS OF ORTHOGONAL COMPONENTS IN REGRESSION IF X IS SPACED AT EQUAL INTERVALS

Degrees of Polynomial	Comparison	Number of Levels							Divisor $\Sigma\lambda^2$
		1	2	3	4	5	6	7	
1	Linear	−1	+1						2
2	Linear	−1	0	+1					2
	Quadratic	+1	−2	+1					6
3	Linear	−3	−1	+1	+3				20
	Quadratic	+1	−1	−1	+1				4
	Cubic	−1	+3	−3	+1				20
4	Linear	−2	−1	0	+1	+2			10
	Quadratic	+2	−1	−2	−1	+2			14
	Cubic	−1	+2	0	−2	+1			10
	Quartic	+1	−4	+6	−4	+1			70
5	Linear	−5	−3	−1	+1	+3	+5		70
	Quadratic	+5	−1	−4	−4	−1	+5		84
	Cubic	−5	+7	+4	−4	−7	+5		180
	Quartic	+1	−3	+2	+2	−3	+1		28
	Quintic	−1	+5	−10	+10	−5	+1		252
6	Linear	−3	−2	−1	0	+1	+2	+3	28
	Quadratic	+5	0	−3	−4	−3	0	+5	84
	Cubic	−1	+1	+1	0	−1	−1	+1	6
	Quartic	+3	−7	+1	+6	+1	−7	+3	154
	Quintic	−1	+4	−5	0	+5	−4	+1	84
	Sextic	+1	−6	+15	−20	+15	−6	+1	924

Source: Snedecor and Cochran (1967).

TABLE G.1. STUDENTIZED RANGE FOR A MULTIPLE-RANGE TEST, FOR p MEANS OR TOTALS AND STANDARD ERRORS WITH n DEGREES OF FREEDOM

5% Protection level.

n \ p	2	3	4	5	6	7	8	9	10	11	12	14	16	18	20	24
1	17.97	17.97	17.97	17.97	17.97	17.97	17.97	17.97	17.97	17.97	17.97	17.97	17.97	17.97	17.97	17.97
2	6.085	6.085	6.085	6.085	6.085	6.085	6.085	6.085	6.085	6.085	6.085	6.085	6.085	6.085	6.085	6.085
3	4.501	4.516	4.516	4.516	4.516	4.516	4.516	4.516	4.516	4.516	4.516	4.516	4.516	4.516	4.516	4.516
4	3.927	4.013	4.033	4.033	4.033	4.033	4.033	4.033	4.033	4.033	4.033	4.033	4.033	4.033	4.033	4.033
5	3.635	3.749	3.797	3.814	3.814	3.814	3.814	3.814	3.814	3.814	3.814	3.814	3.814	3.814	3.814	3.814
6	3.461	3.587	3.649	3.680	3.694	3.697	3.697	3.697	3.697	3.697	3.697	3.697	3.697	3.697	3.697	3.697
7	3.344	3.477	3.548	3.588	3.611	3.622	3.626	3.626	3.626	3.626	3.626	3.626	3.626	3.626	3.626	3.626
8	3.261	3.399	3.475	3.521	3.549	3.566	3.575	3.579	3.579	3.579	3.579	3.579	3.579	3.579	3.579	3.579
9	3.199	3.339	3.420	3.470	3.502	3.523	3.536	3.544	3.547	3.547	3.547	3.547	3.547	3.547	3.547	3.547
10	3.151	3.293	3.376	3.430	3.465	3.489	3.505	3.516	3.522	3.525	3.526	3.526	3.526	3.526	3.526	3.526
11	3.113	3.256	3.342	3.397	3.435	3.462	3.480	3.493	3.501	3.506	3.509	3.510	3.510	3.510	3.510	3.510
12	3.082	3.225	3.313	3.370	3.410	3.439	3.459	3.474	3.484	3.491	3.496	3.499	3.499	3.499	3.499	3.499
13	3.055	3.200	3.289	3.348	3.389	3.419	3.442	3.458	3.470	3.478	3.484	3.490	3.490	3.490	3.490	3.490
14	3.033	3.178	3.268	3.329	3.372	3.403	3.426	3.444	3.457	3.467	3.474	3.482	3.484	3.485	3.485	3.485
15	3.014	3.160	3.250	3.312	3.356	3.389	3.413	3.432	3.446	3.457	3.465	3.476	3.480	3.481	3.481	3.481
16	2.998	3.144	3.235	3.298	3.343	3.376	3.402	3.422	3.437	3.449	3.458	3.470	3.477	3.478	3.478	3.478
17	2.984	3.130	3.222	3.285	3.331	3.366	3.392	3.412	3.429	3.441	3.451	3.465	3.473	3.476	3.476	3.476
18	2.971	3.118	3.210	3.274	3.321	3.356	3.383	3.405	3.421	3.435	3.445	3.460	3.470	3.474	3.474	3.474
20	2.950	3.097	3.190	3.255	3.303	3.339	3.368	3.391	3.409	3.424	3.436	3.453	3.464	3.470	3.473	3.474
24	2.919	3.066	3.160	3.226	3.276	3.315	3.345	3.370	3.390	3.406	3.420	3.441	3.456	3.465	3.471	3.477
30	2.888	3.035	3.131	3.199	3.250	3.290	3.322	3.349	3.371	3.389	3.405	3.430	3.447	3.460	3.470	3.481
40	2.858	3.006	3.102	3.171	3.224	3.266	3.300	3.328	3.352	3.373	3.390	3.418	3.439	3.456	3.469	3.486
60	2.829	2.976	3.073	3.143	3.198	3.241	3.277	3.307	3.333	3.355	3.374	3.406	3.431	3.451	3.467	3.492
120	2.800	2.947	3.045	3.116	3.172	3.217	3.254	3.287	3.314	3.337	3.359	3.394	3.423	3.446	3.466	3.498
∞	2.772	2.918	3.017	3.089	3.146	3.193	3.232	3.265	3.294	3.320	3.343	3.382	3.414	3.442	3.466	3.505

Source: Bliss (1967).

TABLE H.1. VALUES OF χ^2

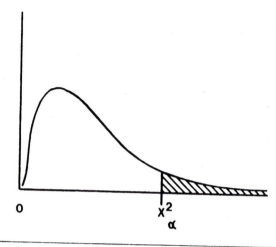

df	Probability of a Larger Value of χ^2				
	0.995	0.990	0.975	0.950	0.900
1	0.0^4393	0.0^3157	0.0^3982	0.0^2393	0.0158
2	0.0100	0.0201	0.0506	0.103	0.211
3	0.0717	0.115	0.216	0.352	0.584
4	0.207	0.297	0.484	0.711	1.06
5	0.412	0.554	0.831	1.15	1.61
6	0.676	0.872	1.24	1.64	2.20
7	0.989	1.24	1.69	2.17	2.83
8	1.34	1.65	2.18	2.73	3.49
9	1.73	2.09	2.70	3.33	4.17
10	2.16	2.56	3.25	3.94	4.87
11	2.60	3.05	3.82	4.57	5.58
12	3.07	3.57	4.40	5.23	6.30
13	3.57	4.11	5.01	5.89	7.04
14	4.07	4.66	5.63	6.57	7.79
15	4.60	5.23	6.26	7.26	8.55
16	5.14	5.81	6.91	7.96	9.31
17	5.70	6.41	7.56	8.67	10.1
18	6.26	7.01	8.23	9.39	10.9
19	6.84	7.63	8.91	10.1	11.7
20	7.43	8.26	9.59	10.9	12.4
21	8.03	8.90	10.3	11.6	13.2
22	8.64	9.54	11.0	12.3	14.0
23	9.26	10.2	11.7	13.1	14.8
24	9.89	10.9	12.4	13.8	15.7
25	10.5	11.5	13.1	14.6	16.5
26	11.2	12.2	13.8	15.4	17.3
27	11.8	12.9	14.6	16.2	18.1
28	12.5	13.6	15.3	16.9	18.9
29	13.1	14.3	16.0	17.7	19.8
30	13.8	15.0	16.8	18.5	20.6
40	20.7	22.2	24.4	26.5	29.1
50	28.0	29.7	32.4	34.8	37.7
60	35.5	37.5	40.5	43.2	46.5

Source: Thompson (1941). (Abridged table reproduced with kind

Probability of a Larger Value of χ^2							
0.750	0.500	0.250	0.100	0.050	0.025	0.010	0.005
0.102	0.455	1.32	2.71	3.84	5.02	6.63	7.88
0.575	1.39	2.77	4.61	5.99	7.38	9.21	10.6
1.21	2.37	4.11	6.25	7.81	9.35	11.3	12.8
1.92	3.36	5.39	7.78	9.49	11.1	13.3	14.9
2.67	4.35	6.63	9.24	11.1	12.8	15.1	16.7
3.45	5.35	7.84	10.6	12.6	14.4	16.8	18.5
4.25	6.35	9.04	12.0	14.1	16.0	18.5	20.3
5.07	7.34	10.2	13.4	15.5	17.5	20.1	22.0
5.90	8.34	11.4	14.7	16.9	19.0	21.7	23.6
6.74	9.34	12.5	16.0	18.3	20.5	23.2	25.2
7.58	10.3	13.7	17.3	19.7	21.9	24.7	26.8
8.44	11.3	14.8	18.5	21.0	23.3	26.2	28.3
9.30	12.3	16.0	19.8	22.4	24.7	27.7	29.8
10.2	13.3	17.1	21.1	23.7	26.1	29.1	31.3
11.0	14.3	18.2	22.3	25.0	27.5	30.6	32.8
11.9	15.3	19.4	23.5	26.3	28.8	32.0	34.3
12.8	16.3	20.5	24.8	27.6	30.2	33.4	35.7
13.7	17.3	21.6	26.0	28.9	31.5	34.8	37.2
14.6	18.3	22.7	27.2	30.1	32.9	36.2	38.6
15.5	19.3	23.8	28.4	31.4	34.2	37.6	40.0
16.3	20.3	24.9	29.6	32.7	35.5	38.9	41.4
17.2	21.3	26.0	30.8	33.9	36.8	40.3	42.8
18.1	22.3	27.1	32.0	35.2	38.1	41.6	44.2
19.0	23.3	28.2	33.2	36.4	39.4	43.0	45.6
19.9	24.3	29.3	34.4	37.7	40.6	44.3	46.9
20.8	25.3	30.4	35.6	38.9	41.9	45.6	48.3
21.7	26.3	31.5	36.7	40.1	43.2	47.0	49.6
22.7	27.3	32.6	37.9	41.3	44.5	48.3	51.0
23.6	28.3	33.7	39.1	42.6	45.7	49.6	52.3
24.5	29.3	34.8	40.3	43.8	47.0	50.9	53.7
33.7	39.3	45.6	51.8	55.8	59.3	63.7	66.8
42.9	49.3	56.3	63.2	67.5	71.4	76.2	79.5
52.3	59.3	67.0	74.4	79.1	83.3	88.4	92.0

TABLE I.1. KRAMER RANK SUM 5%

Rank totals required for significance at the 5% level (P≤0.05). The 4 figure blocks represent: lowest insignificant rank sum, any treatment–highest insignificant rank sum, any treatment; lowest insignificant rank sum, predetermined treatment–highest insignificant rank sum, predetermined treatment.

No. of Replications	Number of Treatments, or Samples Ranked																		
	2	3	4	5	6	7	8	9	10	11	12	13	14	15	16	17	18	19	20
2	—	—	—	3-9	3-11	3-13	4-14	4-16	4-18	5-19	5-21	5-23	5-25	6-26	6-28	6-30	7-31	7-33	3-39 / 7-35
3	—	4-8	4-11	4-14 / 5-13	4-17 / 6-15	4-20 / 6-18	4-23 / 7-20	5-25 / 8-22	5-28 / 8-25	5-31 / 9-27	5-34 / 10-29	5-37 / 10-32	5-40 / 11-34	6-42 / 12-36	6-45 / 12-39	6-48 / 13-41	6-51 / 14-43	6-54 / 14-46	7-56 / 15-48
4	—	5-11 / 5-11	5-15 / 6-14	6-18 / 7-17	6-21 / 8-20	7-25 / 9-23	7-29 / 10-26	8-32 / 11-29	8-36 / 13-31	8-40 / 14-34	9-43 / 15-37	9-47 / 16-40	10-50 / 17-43	10-54 / 18-46	10-58 / 19-49	11-61 / 20-52	11-65 / 21-55	12-68 / 22-58	12-72 / 23-61
5	6-9 / 6-9	6-14 / 7-13	7-18 / 8-17	8-22 / 10-20	9-26 / 11-24	9-31 / 13-27	10-35 / 14-31	11-39 / 15-35	12-43 / 17-38	12-48 / 18-42	13-52 / 20-45	14-56 / 21-49	14-61 / 23-52	15-65 / 24-56	16-69 / 25-60	16-74 / 27-63	17-78 / 28-67	18-82 / 30-70	18-87 / 31-74
6	7-11 / 7-11	8-16 / 9-15	9-21 / 11-19	10-26 / 12-24	11-31 / 14-28	12-36 / 16-32	13-41 / 18-36	14-46 / 20-40	15-51 / 21-45	17-55 / 23-49	18-60 / 25-53	19-65 / 27-57	19-71 / 29-61	20-76 / 31-65	21-81 / 32-70	22-86 / 34-74	23-91 / 36-78	24-96 / 38-82	25-101 / 40-86
7	8-13 / 8-13	10-18 / 10-18	11-24 / 13-22	12-30 / 15-27	14-35 / 17-32	15-41 / 19-37	17-46 / 22-41	18-52 / 26-46	19-58 / 26-51	21-63 / 28-56	22-69 / 30-61	23-75 / 33-65	25-80 / 35-70	26-86 / 37-75	27-92 / 39-80	29-97 / 42-84	30-103 / 44-89	31-109 / 46-94	32-115 / 48-99
8	9-15 / 10-14	11-21 / 12-20	13-27 / 15-25	15-33 / 17-31	17-39 / 20-36	18-46 / 23-41	20-52 / 25-47	22-58 / 28-52	24-64 / 31-57	25-71 / 33-63	27-77 / 36-68	29-83 / 39-73	30-90 / 41-79	32-96 / 44-84	33-103 / 47-89	35-109 / 49-95	37-115 / 52-100	38-122 / 54-106	40-128 / 57-111
9	11-16 / 11-16	13-23 / 14-22	15-30 / 17-28	17-37 / 20-34	19-44 / 23-40	22-50 / 26-46	24-57 / 29-52	26-64 / 32-58	28-71 / 35-64	30-78 / 38-70	32-85 / 41-76	34-92 / 45-81	36-99 / 48-87	38-106 / 51-93	40-113 / 54-99	42-120 / 57-105	44-127 / 60-111	45-135 / 63-117	47-142 / 66-123
10	12-18 / 12-18	15-25 / 16-24	17-33 / 19-31	20-40 / 23-37	22-48 / 26-44	25-55 / 30-50	27-63 / 33-57	30-70 / 37-63	32-78 / 40-70	34-86 / 44-76	37-93 / 47-83	39-101 / 51-89	41-109 / 54-96	44-116 / 57-103	46-124 / 61-109	48-132 / 64-116	51-139 / 68-122	53-147 / 71-129	55-155 / 75-135
11	13-20 / 14-19	16-28 / 18-26	19-36 / 21-34	22-44 / 25-41	25-52 / 29-48	28-60 / 33-55	31-68 / 37-62	34-76 / 41-69	36-85 / 45-76	39-93 / 49-83	42-101 / 53-90	45-109 / 57-97	47-118 / 60-105	50-126 / 64-112	53-134 / 68-119	55-143 / 72-126	58-151 / 76-133	60-160 / 80-140	63-168 / 84-147
12	15-21 / 15-21	18-30 / 19-29	21-39 / 23-36	25-47 / 28-44	28-56 / 32-52	31-65 / 37-59	34-74 / 41-67	38-82 / 45-75	41-91 / 50-82	44-100 / 54-90	47-109 / 58-98	50-118 / 63-105	53-127 / 67-113	56-136 / 71-121	59-145 / 76-128	62-154 / 80-136	65-163 / 84-144	68-172 / 89-151	71-181 / 93-159
13	16-23 / 17-22	20-32 / 21-31	23-42 / 25-39	27-51 / 31-47	31-60 / 35-56	35-69 / 40-64	38-79 / 45-72	42-88 / 50-80	45-98 / 54-89	49-107 / 59-97	52-117 / 64-105	56-126 / 69-113	59-136 / 74-121	62-146 / 78-130	66-155 / 83-138	69-165 / 88-146	73-174 / 93-154	76-184 / 97-163	79-194 / 102-171
14	17-25 / 18-24	22-34 / 23-33	26-44 / 28-42	30-54 / 33-51	34-64 / 38-60	38-74 / 44-68	42-84 / 49-77	46-94 / 54-86	50-104 / 59-95	54-114 / 65-103	57-125 / 70-112	61-135 / 75-121	65-145 / 80-130	69-155 / 85-139	73-165 / 91-147	75-176 / 96-156	80-186 / 101-165	84-196 / 106-174	88-206 / 111-183
15	19-26 / 19-26	23-37 / 25-35	28-47 / 30-45	32-58 / 36-54	37-68 / 42-64	41-79 / 47-73	46-89 / 53-82	50-100 / 59-91	54-111 / 64-101	60-122 / 70-110	63-132 / 75-120	67-143 / 81-129	71-154 / 87-138	75-165 / 92-148	79-176 / 98-157	82-187 / 104-166	88-197 / 109-176	92-208 / 115-185	96-219 / 121-194
16	20-28 / 21-27	25-39 / 27-37	30-50 / 33-47	35-61 / 39-57	40-72 / 45-67	45-83 / 51-77	49-95 / 57-87	54-106 / 63-97	59-117 / 69-107	65-129 / 75-117	68-140 / 81-127	72-151 / 87-137	77-163 / 93-147	82-174 / 100-156	86-186 / 106-166	89-197 / 112-176	95-209 / 118-186	100-220 / 124-196	104-232 / 130-206
17	22-29 / 22-29	27-40 / 28-40	32-53 / 35-50	37-64 / 41-61	43-76 / 48-71	48-88 / 54-82	53-100 / 61-92	58-112 / 67-103	63-124 / 73-113	70-136 / 81-123	73-146 / 87-134	78-160 / 94-144	83-172 / 100-155	88-184 / 107-165	93-196 / 113-176	95-208 / 118-186	103-220 / 126-197	108-232 / 133-207	113-244 / 139-218
18	23-31 / 24-30	29-43 / 30-42	34-56 / 37-53	40-68 / 44-64	46-80 / 51-75	51-93 / 58-86	57-105 / 65-97	62-118 / 72-108	68-130 / 78-119	76-143 / 87-134	79-155 / 93-141	84-168 / 100-152	90-180 / 107-163	95-193 / 114-174	100-206 / 121-185	102-218 / 126-196	111-231 / 135-207	116-244 / 142-218	121-257 / 148-229
19	24-33 / 25-32	30-46 / 32-44	37-59 / 39-57	42-71 / 47-67	49-84 / 54-79	55-97 / 62-90	61-110 / 69-102	67-123 / 76-114	73-136 / 83-125	81-150 / 93-141	84-163 / 99-148	90-176 / 106-160	96-189 / 114-171	102-202 / 121-183	107-216 / 128-195	109-229 / 136-206	119-242 / 143-218	124-256 / 151-229	130-269 / 158-241

TABLE J1.   KRAMER RANK SUM 1%

Rank totals required for significance at the 1% level (P≤0.01). The 4 figure blocks represent: lowest insignificant rank sum, any treatment–highest insignificant rank sum, any treatment; lowest insignificant rank sum, predetermined treatment–highest insignificant rank sum, predetermined treatment.

Number of Treatments, or Samples Ranked

No. of Replications	2	3	4	5	6	7	8	9	10	11	12	13	14	15	16	17	18	19	20
2	—	—	—	—	—	—	—	—	3-19	3-21	3-23	3-25	3-27	3-29	3-31	3-33	4-34	4-36	4-36
3	—	—	—	4-14	4-17	4-20	5-22	5-25	4-29 / 6-27	4-32 / 6-30	4-35 / 6-33	4-38 / 7-35	4-41 / 7-38	4-44 / 7-41	4-47 / 8-43	4-50 / 8-46	4-53 / 9-48	4-56 / 9-51	4-59 / 9-54
4	—	—	5-15	5-19 / 6-18	5-23 / 6-22	5-27 / 7-25	6-30 / 8-28	6-34 / 8-32	6-38 / 9-35	6-42 / 10-38	7-45 / 10-42	7-49 / 11-45	7-53 / 12-48	7-57 / 13-51	8-60 / 13-55	8-64 / 14-58	9-68 / 15-62	8-72 / 15-65	9-75 / 16-68
5	—	6-14	6-19 / 7-18	7-23 / 8-22	7-28 / 9-26	8-32 / 10-30	8-37 / 11-34	9-41 / 12-38	9-46 / 13-42	10-50 / 14-46	10-55 / 15-50	11-59 / 16-54	11-64 / 16-59	12-68 / 18-62	12-73 / 19-66	13-77 / 20-70	13-82 / 21-74	14-86 / 22-78	14-91 / 23-82
6	—	7-17 / 8-16	8-22 / 9-21	9-27 / 10-26	9-33 / 12-30	10-38 / 13-35	11-43 / 14-40	12-48 / 16-44	13-53 / 17-49	14-58 / 18-54	14-64 / 20-58	15-69 / 21-63	16-74 / 23-67	16-80 / 24-72	17-85 / 25-77	18-90 / 27-81	19-95 / 28-86	19-101 / 29-91	20-106 / 31-95
7	8-13	8-20 / 9-19	10-25 / 11-24	11-31 / 12-30	12-37 / 14-35	13-43 / 16-40	14-49 / 18-45	15-55 / 19-51	16-61 / 21-56	17-67 / 23-61	18-73 / 25-66	19-79 / 26-72	21-85 / 28-77	21-91 / 30-82	22-97 / 32-87	23-103 / 33-93	24-109 / 35-98	25-115 / 37-103	26-121 / 39-108
8	9-15 / 9-15	10-22 / 11-21	11-29 / 13-27	13-35 / 15-33	14-42 / 17-39	16-48 / 19-45	17-55 / 21-51	19-61 / 23-57	20-68 / 25-63	21-75 / 28-68	23-81 / 30-74	24-88 / 32-80	26-95 / 34-86	27-101 / 36-92	28-108 / 38-98	29-115 / 40-104	31-121 / 43-109	32-128 / 45-115	33-135 / 47-121
9	10-17 / 10-17	12-24 / 12-24	13-32 / 15-30	15-39 / 17-37	17-46 / 20-43	19-53 / 22-50	21-60 / 25-56	22-68 / 27-63	24-75 / 30-69	26-82 / 32-76	27-90 / 35-82	29-97 / 37-89	31-104 / 40-95	32-112 / 42-102	34-119 / 45-108	35-127 / 47-115	37-134 / 50-121	39-141 / 52-128	40-149 / 55-134
10	11-19 / 11-19	13-27 / 14-26	15-35 / 17-33	18-42 / 20-40	20-50 / 23-47	22-58 / 25-55	24-66 / 28-62	26-74 / 31-69	28-82 / 34-76	30-90 / 37-83	32-98 / 40-90	34-106 / 43-97	36-114 / 46-104	38-122 / 49-111	40-130 / 52-118	42-138 / 54-126	44-146 / 57-133	46-154 / 60-140	48-162 / 63-147
11	12-21 / 13-20	15-29 / 16-28	17-38 / 19-36	20-46 / 22-44	22-55 / 25-52	25-63 / 29-59	27-72 / 32-67	30-80 / 35-75	32-89 / 39-82	34-98 / 42-90	37-106 / 45-98	39-115 / 48-106	41-124 / 52-114	44-132 / 55-121	46-141 / 58-129	48-150 / 61-137	50-159 / 65-144	53-167 / 68-152	55-176 / 72-159
12	14-22 / 14-22	16-31 / 17-30	19-41 / 21-39	22-50 / 25-47	25-59 / 28-56	28-68 / 32-64	31-77 / 36-72	33-87 / 39-81	36-96 / 43-89	39-105 / 47-97	42-114 / 50-106	44-124 / 54-114	47-133 / 58-122	50-142 / 62-130	52-152 / 65-139	55-161 / 68-147	57-171 / 72-156	60-180 / 76-164	63-189 / 80-172
13	15-24 / 15-24	18-34 / 19-33	21-44 / 23-42	25-53 / 27-51	28-63 / 31-60	31-73 / 35-69	34-83 / 39-78	37-93 / 44-86	40-103 / 48-95	43-113 / 52-104	46-123 / 56-113	50-132 / 60-122	53-142 / 64-131	56-152 / 68-140	58-163 / 72-149	61-173 / 76-158	64-183 / 80-167	67-193 / 85-175	70-203 / 89-184
14	16-26 / 17-25	20-36 / 21-35	24-46 / 25-45	27-57 / 30-54	31-67 / 34-64	34-79 / 39-73	38-88 / 43-83	41-99 / 48-92	45-109 / 52-102	48-120 / 57-111	51-131 / 61-121	55-141 / 66-130	58-152 / 70-140	62-162 / 75-149	65-173 / 79-159	68-184 / 84-168	71-195 / 88-178	75-205 / 93-187	78-216 / 97-197
15	18-27 / 18-27	22-38 / 23-37	26-49 / 28-47	30-60 / 32-58	34-71 / 37-68	37-83 / 42-78	41-94 / 47-88	45-105 / 52-98	49-116 / 57-108	53-127 / 62-118	56-139 / 67-128	60-150 / 72-138	64-161 / 76-149	68-172 / 81-159	71-184 / 86-169	75-195 / 91-179	79-206 / 96-189	82-218 / 101-199	86-229 / 106-209
16	19-29 / 19-29	23-41 / 25-39	28-52 / 30-50	32-64 / 35-61	36-76 / 40-72	41-87 / 46-82	45-99 / 51-93	49-111 / 56-104	53-123 / 61-115	57-135 / 67-125	62-146 / 72-136	66-158 / 77-147	70-170 / 83-157	74-182 / 88-168	78-194 / 93-179	82-206 / 99-189	86-218 / 104-200	90-230 / 109-211	94-242 / 115-221
17	20-31 / 21-30	25-43 / 26-42	30-55 / 32-53	35-67 / 38-64	39-80 / 43-76	44-92 / 50-87	49-105 / 55-98	53-117 / 60-110	58-129 / 66-121	62-142 / 72-132	67-154 / 78-143	71-167 / 83-155	76-179 / 89-166	80-192 / 95-177	85-204 / 101-188	89-217 / 106-200	93-230 / 112-211	98-242 / 118-222	102-255 / 124-233
18	22-32 / 22-32	27-45 / 28-44	32-58 / 34-56	37-71 / 40-68	42-84 / 46-80	47-97 / 53-92	52-110 / 59-103	57-123 / 65-115	62-136 / 71-127	67-149 / 77-139	72-162 / 83-151	77-175 / 89-163	82-188 / 95-175	86-202 / 102-186	91-215 / 108-200	96-228 / 114-210	101-241 / 120-222	106-254 / 126-234	110-268 / 132-246
19	23-34 / 24-33	29-47 / 30-46	34-61 / 36-59	40-74 / 43-71	45-88 / 49-84	50-102 / 56-96	56-115 / 62-109	61-129 / 69-121	67-142 / 76-133	72-156 / 82-146	77-170 / 89-158	82-184 / 95-171	87-197 / 102-183	93-211 / 108-196	98-225 / 115-208	103-239 / 122-220	108-253 / 128-233	113-267 / 135-245	119-280 / 141-258

REFERENCES

BLISS, G.I.   1967.   Statistics in Biology.   McGraw-Hill Book Co., New York.

FISHER, R.A.   1925.   Statistical Methods for Research Workers.   Oliver and Boyd, Edinburgh.   Reproduced by permission of the author and publishers.

HAMBURG, M.   1970.   Statistical Analysis for Decision Making.   Harcourt Brace Jovanovich, New York.   Reproduced with the permission of the publishers.

HICKMAN, E.P. and HILTON, J.G.   1971.   Probability and Statistical Analysis.   Intext Educational Publishers, Scranton, Penn.

SNEDECOR, G.W. and COCHRAN, W.G.   1967.   Statistical Methods, 6th Edition.   Iowa State Univ. Press, Ames.

THOMPSON, C.M.   1941.   Table of percentage points of the χ^2 distribution.   Biometrika 32, 188–189.   Abridged table published with kind permission of the author and the editor.

TEMPERATURE CONVERSION

The numbers in boldface type in the center column refer to the temperature, either in degrees Celsius or Fahrenheit, which is to be converted to the other scale. If converting Fahrenheit to degree Celsius, the equivalent temperature will be found in the left column. If converting degree Celsius to Fahrenheit, the equivalent temperature will be found in the column on the right.

Temperature			Temperature			Temperature			Temperature		
Celsius	°C or F	Fahr	Celsius	°C or F	Fahr	Celsius	°C or F	Fahr	Celsius	°C or F	Fahr
-40.0	-40	-40.0	+1.7	+35	+95.0	+43.3	+110	+230.0	+85.0	+185	+365.0
-39.4	-39	-38.2	+2.2	+36	+96.8	+43.9	+111	+231.8	+85.6	+186	+366.8
-38.9	-38	-36.4	+2.8	+37	+98.6	+44.4	+112	+233.6	+86.1	+187	+368.6
-38.3	-37	-34.6	+3.3	+38	+100.4	+45.0	+113	+235.4	+86.7	+188	+370.4
-37.8	-36	-32.8	+3.9	+39	+102.2	+45.6	+114	+237.2	+87.2	+189	+372.2
-37.2	-35	-31.0	+4.4	+40	+104.0	+46.1	+115	+239.0	+87.8	+190	+374.0
-36.7	-34	-29.2	+5.0	+41	+105.8	+46.7	+116	+240.8	+88.3	+191	+375.8
-36.1	-33	-27.4	+5.5	+42	+107.6	+47.2	+117	+242.6	+88.9	+192	+377.6
-35.6	-32	-25.6	+6.1	+43	+109.4	+47.8	+118	+244.4	+89.4	+193	+379.4
-35.0	-31	-23.8	+6.7	+44	+111.2	+48.3	+119	+246.2	+90.0	+194	+381.2
-34.4	-30	-22.0	+7.2	+45	+113.0	+48.9	+120	+248.0	+90.6	+195	+383.0
-33.9	-29	-20.2	+7.8	+46	+114.8	+49.4	+121	+249.8	+91.1	+196	+384.8
-33.3	-28	-18.4	+8.3	+47	+116.6	+50.0	+122	+251.6	+91.7	+197	+386.6
-32.8	-27	-16.6	+8.9	+48	+118.4	+50.6	+123	+253.4	+92.2	+198	+388.4
-32.2	-26	-14.8	+9.4	+49	+120.2	+51.1	+124	+255.2	+92.8	+199	+390.2
-31.7	-25	-13.0	+10.0	+50	+122.0	+51.7	+125	+257.0	+93.3	+200	+392.0
-31.1	-24	-11.2	+10.6	+51	+123.8	+52.2	+126	+258.8	+93.9	+201	+393.8
-30.6	-23	-9.4	+11.1	+52	+125.6	+52.8	+127	+260.6	+94.4	+202	+395.6
-30.0	-22	-7.6	+11.7	+53	+127.4	+53.3	+128	+262.4	+95.0	+203	+397.4
-29.4	-21	-5.8	+12.2	+54	+129.2	+53.9	+129	+264.2	+95.6	+204	+399.2
-28.9	-20	-4.0	+12.8	+55	+131.0	+54.4	+130	+266.0	+96.1	+205	+401.0
-28.3	-19	-2.2	+13.3	+56	+132.8	+55.0	+131	+267.8	+96.7	+206	+402.8
-27.8	-18	-0.4	+13.9	+57	+134.6	+55.6	+132	+269.6	+97.2	+207	+404.6
-27.2	-17	+1.4	+14.4	+58	+136.4	+56.1	+133	+271.4	+97.8	+208	+406.4
-26.7	-16	+3.2	+15.0	+59	+138.2	+56.7	+134	+273.2	+98.3	+209	+408.2
-26.1	-15	+5.0	+15.6	+60	+140.0	+57.2	+135	+275.0	+98.9	+210	+410.0
-25.6	-14	+6.8	+16.1	+61	+141.8	+57.8	+136	+276.8	+99.4	+211	+411.8
-25.0	-13	+8.6	+16.7	+62	+143.6	+58.3	+137	+278.6	+100.0	+212	+413.6
-24.4	-12	+10.4	+17.2	+63	+145.4	+58.9	+138	+280.4	+100.6	+213	+415.4
-23.9	-11	+12.2	+17.8	+64	+147.2	+59.4	+139	+282.2	+101.1	+214	+417.2
-23.3	-10	+14.0	+18.3	+65	+149.0	+60.0	+140	+284.0	+101.7	+215	+419.0
-22.8	-9	+15.8	+18.9	+66	+150.8	+60.6	+141	+285.8	+102.2	+216	+420.8
-22.2	-8	+17.6	+19.4	+67	+152.6	+61.1	+142	+287.6	+102.8	+217	+422.6
-21.7	-7	+19.4	+20.0	+68	+154.4	+61.7	+143	+289.4	+103.3	+218	+424.4
-21.1	-6	+21.2	+20.6	+69	+156.2	+62.2	+144	+291.2	+103.9	+219	+426.2
-20.6	-5	+23.0	+21.1	+70	+158.0	+62.8	+145	+293.0	+104.4	+220	+428.0
-20.0	-4	+24.8	+21.7	+71	+159.8	+63.3	+146	+294.8	+105.6	+222	+431.6
-19.4	-3	+26.6	+22.2	+72	+161.6	+63.9	+147	+296.6	+106.7	+224	+435.2
-18.9	-2	+28.4	+22.8	+73	+163.4	+64.4	+148	+298.4	+107.8	+226	+438.8
-18.3	-1	+30.2	+23.3	+74	+165.2	+65.0	+149	+300.2	+108.9	+228	+442.4
-17.8	0	+32.0	+23.9	+75	+167.0	+65.6	+150	+302.0	+110.0	+230	+446.0
-17.2	+1	+33.8	+24.4	+76	+168.8	+66.1	+151	+303.8	+111.1	+232	+449.6
-16.7	+2	+35.6	+25.0	+77	+170.6	+66.7	+152	+305.6	+112.2	+234	+453.2
-16.1	+3	+37.4	+25.6	+78	+172.4	+67.2	+153	+307.4	+113.3	+236	+456.8
-15.6	+4	+39.2	+26.1	+79	+174.2	+67.8	+154	+309.2	+114.4	+238	+460.4
-15.0	+5	+41.0	+26.7	+80	+176.0	+68.3	+155	+311.0	+115.6	+240	+464.0
-14.4	+6	+42.8	+27.2	+81	+177.8	+68.9	+156	+312.8	+116.7	+242	+467.6
-13.9	+7	+44.6	+27.8	+82	+179.6	+69.4	+157	+314.6	+117.8	+244	+471.2
-13.3	+8	+46.4	+28.3	+83	+181.4	+70.0	+158	+316.4	+118.9	+246	+474.2
-12.8	+9	+48.2	+28.9	+84	+183.2	+70.6	+159	+318.2	+120.0	+248	+478.4
-12.2	+10	+50.0	+29.4	+85	+185.0	+71.1	+160	+320.0	+121.1	+250	+482.0
-11.7	+11	+51.8	+30.0	+86	+186.8	+71.7	+161	+321.8	+122.4	+252	+485.6
-11.1	+12	+53.6	+30.6	+87	+188.6	+72.2	+162	+323.6	+123.3	+254	+489.2
-10.6	+13	+55.4	+31.1	+88	+190.4	+72.8	+163	+325.4	+124.4	+256	+492.8
-10.0	+14	+57.2	+31.7	+89	+192.2	+73.3	+164	+327.2	+125.5	+258	+496.4
-9.4	+15	+59.0	+32.2	+90	+194.0	+73.9	+165	+329.0	+126.7	+260	+500.0
-8.9	+16	+60.8	+32.8	+91	+195.8	+74.4	+166	+330.8	+127.8	+262	+503.6
-8.3	+17	+62.6	+33.3	+92	+197.6	+75.0	+167	+332.6	+128.9	+264	+507.2
-7.8	+18	+64.4	+33.9	+93	+199.4	+75.6	+168	+334.4	+130.0	+266	+510.8
-7.2	+19	+66.2	+34.4	+94	+201.2	+76.1	+169	+336.2	+131.3	+268	+514.4
-6.7	+20	+68.0	+35.0	+95	+203.0	+76.7	+170	+338.0	+132.2	+270	+518.0
-6.1	+21	+69.8	+35.6	+96	+204.8	+77.2	+171	+339.8	+133.3	+272	+521.6
-5.5	+22	+71.6	+36.1	+97	+206.6	+77.8	+172	+341.6	+134.4	+274	+525.2
-5.0	+23	+73.4	+36.7	+98	+208.4	+78.3	+173	+343.4	+135.6	+276	+528.8
-4.4	+24	+75.2	+37.2	+99	+210.2	+78.9	+174	+345.2	+136.7	+278	+532.4
-3.9	+25	+77.0	+37.8	+100	+212.0	+79.4	+175	+347.0	+137.8	+280	+536.0
-3.3	+26	+78.8	+38.3	+101	+213.8	+80.0	+176	+348.8	+138.9	+282	+539.6
-2.8	+27	+80.6	+38.9	+102	+215.6	+80.6	+177	+350.6	+140.0	+284	+543.2
-2.2	+28	+82.4	+39.4	+103	+217.4	+81.1	+178	+352.4	+141.1	+286	+546.8
-1.7	+29	+84.2	+40.0	+104	+219.2	+81.7	+179	+354.2	+142.2	+288	+550.4
-1.1	+30	+86.0	+40.6	+105	+221.0	+82.2	+180	+356.0	+143.3	+290	+554.0
-0.6	+31	+87.8	+41.1	+106	+222.8	+82.8	+181	+357.8	+144.4	+292	+557.6
.0	+32	+89.6	+41.7	+107	+224.6	+83.3	+182	+359.6	+145.6	+294	+561.2
+0.6	+33	+91.4	+42.2	+108	+226.4	+83.9	+183	+361.4	+146.7	+296	+564.8
+1.1°	+34	+93.2	+42.8	+109	+228.2	+84.4	+184	+363.2	+147.8	+298	+568.4

COMPARISON OF AVOIRDUPOIS AND METRIC UNITS OF WEIGHT

1 oz = 0.06 lb = 28.35 g	1 lb = 0.454 kg	1 g = 0.035 oz	1 kg = 2.205 lb
2 oz = 0.12 lb = 56.70 g	2 lb = 0.91 kg	2 g = 0.07 oz	2 kg = 4.41 lb
3 oz = 0.19 lb = 85.05 g	3 lb = 1.36 kg	3 g = 0.11 oz	3 kg = 6.61 lb
4 oz = 0.25 lb = 113.40 g	4 lb = 1.81 kg	4 g = 0.14 oz	4 kg = 8.82 lb
5 oz = 0.31 lb = 141.75 g	5 lb = 2.27 kg	5 g = 0.18 oz	5 kg = 11.02 lb
6 oz = 0.38 lb = 170.10 g	6 lb = 2.72 kg	6 g = 0.21 oz	6 kg = 13.23 lb
7 oz = 0.44 lb = 198.45 g	7 lb = 3.18 kg	7 g = 0.25 oz	7 kg = 15.43 lb
8 oz = 0.50 lb = 226.80 g	8 lb = 3.63 kg	8 g = 0.28 oz	8 kg = 17.64 lb
9 oz = 0.56 lb = 255.15 g	9 lb = 4.08 kg	9 g = 0.32 oz	9 kg = 19.84 lb
10 oz = 0.62 lb = 283.50 g	10 lb = 4.54 kg	10 g = 0.35 oz	10 kg = 22.05 lb
11 oz = 0.69 lb = 311.85 g	11 lb = 4.99 kg	11 g = 0.39 oz	11 kg = 24.26 lb
12 oz = 0.75 lb = 340.20 g	12 lb = 5.44 kg	12 g = 0.42 oz	12 kg = 26.46 lb
13 oz = 0.81 lb = 368.55 g	13 lb = 5.90 kg	13 g = 0.46 oz	13 kg = 28.67 lb
14 oz = 0.88 lb = 396.90 g	14 lb = 6.35 kg	14 g = 0.49 oz	14 kg = 30.87 lb
15 oz = 0.94 lb = 425.25 g	15 lb = 6.81 kg	15 g = 0.53 oz	15 kg = 33.08 lb
16 oz = 1.00 lb = 453.59 g	16 lb = 7.26 kg	16 g = 0.56 oz	16 kg = 35.28 lb

COMPARISON OF U.S. AND METRIC UNITS OF LIQUID MEASURE

1 fl oz = 29.573 ml	1 qt = 0.946 liter	1 gal. = 3.785 liters
2 fl oz = 59.15 ml	2 qt = 1.89 liters	2 gal. = 7.57 liters
3 fl oz = 88.72 ml	3 qt = 2.84 liters	3 gal. = 11.36 liters
4 fl oz = 118.30 ml	4 qt = 3.79 liters	4 gal. = 15.14 liters
5 fl oz = 147.87 ml	5 qt = 4.73 liters	5 gal. = 18.93 liters
6 fl oz = 177.44 ml	6 qt = 5.68 liters	6 gal. = 22.71 liters
7 fl oz = 207.02 ml	7 qt = 6.62 liters	7 gal. = 26.50 liters
8 fl oz = 236.59 ml	8 qt = 7.57 liters	8 gal. = 30.28 liters
9 fl oz = 266.16 ml	9 qt = 8.52 liters	9 gal. = 34.07 liters
10 fl oz = 295.73 ml	10 qt = 9.46 liters	10 gal. = 37.85 liters

1 ml = 0.034 fl oz	1 liter = 1.057 qt	1 liter = 0.264 gal.
2 ml = 0.07 fl oz	2 liters = 2.11 qt	2 liters = 0.53 gal.
3 ml = 0.10 fl oz	3 liters = 3.17 qt	3 liters = 0.79 gal.
4 ml = 0.14 fl oz	4 liters = 4.23 qt	4 liters = 1.06 gal.
5 ml = 0.17 fl oz	5 liters = 5.28 qt	5 liters = 1.32 gal.
6 ml = 0.20 fl oz	6 liters = 6.34 qt	6 liters = 1.59 gal.
7 ml = 0.24 fl oz	7 liters = 7.40 qt	7 liters = 1.85 gal.
8 ml = 0.27 fl oz	8 liters = 8.45 qt	8 liters = 2.11 gal.
9 ml = 0.30 fl oz	9 liters = 9.51 qt	9 liters = 2.38 gal.
10 ml = 0.34 fl oz	10 liters = 10.57 qt	10 liters = 2.64 gal.

CONVERSION OF OVEN TEMPERATURES

Conventional (Fahrenheit)		Metric (Celsius)
200 F		93 C
225 F		107 C
250 F	Very low	121 C
300 F	Low	149 C
325 F		163 C
350 F	Moderate	177 C
400 F	Hot	204 C
450 F	Very high	232 C
500 F	Extremely high	260 C

VOLUME CONVERSION DIFFERENCES
CONVENTIONAL VS. METRIC MEASUREMENTS

Utensil	Capacity (ml)	Tolerance (ml)
1 cup	236.6	11.8
½ cup	118.3	5.9
⅓ cup	78.9	3.9
¼ cup	59.2	3.0
1 tablespoon	14.79	0.73
1 teaspoon	4.93	0.24
½ teaspoon	2.46	0.12
¼ teaspoon	1.23	0.06

Index

m780-1-TN

13